ARCO

Everything you need to score high

Math for Smart Test-Takers

SAT · PSAT · ACT · GRE · GMAT

Mark Alan Stewart

Macmillan • USA

Macmillan Reference USA
A Simon & Schuster Macmillan Company
1633 Broadway
New York, NY 10019-6785

An ARCO Book

ARCO is a registered trademark of Simon & Schuster Inc.
MACMILLAN is a registered trademark of Macmillan, Inc.

Manufactured in th United States of America

10 9 8 7 6 5 4 3 2

Library of Congress Number: 97-80925

ISBN: 0-02-862181-6

Contents

1

The Smart Way
to Learn Math

By opening this book, you've made an important first step toward mastering the mathematical concepts you need to know to score high on standardized tests such as the SAT, GRE, ACT, and GMAT. This book takes a unique approach. It's not a textbook. Instead, it "talks" about math in much the same way a real teacher would explain the concepts to you in a live classroom setting. You'll find this approach especially helpful if you're a "verbal" person who suffers from "mathphobia." Nevertheless, test-takers at all levels of mathematical ability can benefit from this book.

Before we get started sharpening your math skills, let's first answer some frequently asked questions about preparing for standardized math tests. We'll also take a sneak peak at what lies ahead in this book and how to use the materials most effectively.

Questions and Answers about Preparing for Standardized Math Exams

Q: *What general areas of math are covered on standardized exams such as the SAT, GRE, ACT, and GMAT?*

A: This book's Table of Contents should give you a good idea about the scope of math topics covered on these exams. There's nothing here (and on the exams) that any high-school student has not already encountered in basic algebra and geometry courses. Advanced topics such as trigonometry and calculus are not covered on the exams.

Q: *Shouldn't I study for the math portion of my particular exam using a book written specifically for that exam?*

A: Not necessarily. The math skills tested are essentially the same on all of the tests (SAT, GRE, ACT, and GMAT). It's the *format* (number of questions and how the problem is posed) that differs among the tests. By all means, familiarize yourself with your exam's format—the math section as well as all other sections. (A myriad of publications with sample tests are available for this purpose.) However, which study guide you should use specifically for math problems depends on how effectively the guide conveys the underlying concepts, *not* on which exam the book is written about.

Q: *I suffer from "mathphobia." Is there any cure?*

A: Probably not. At the risk of overgeneralizing, it seems there are two kinds of people in this world: those who like math, and everyone else. People in the former group are generally good at math "naturally," while the mathphobics (people in the latter group) aren't good at it. Although there's no permanent cure for mathphobia, you don't have to throw in the towel and give up. Work your way through this book, noting which concepts are "clicking" for you. Then focus on those areas, not on the trouble areas that seem to confound you no matter how hard you try to understand. This advice may seem backward, but it's nevertheless good advice for test-takers with weak math skills. Dwelling on trouble areas may only heighten your anxiety, causing you to miss even more of the math questions on your exam. Bolster your confidence by letting go of those trouble spots. Instead, master the concepts you know you're capable of understanding. That way, you'll perform your personal best on the math portion of your test. You can't ask any more of yourself, can you?

Q: *My math skills are already strong, so why should I bother reviewing basic math concepts?*

A: If you think you can waltz into a standardized test session and come away with a perfect Quantitative score, get ready to eat some humble pie! Sure, if you're a "math wiz" the concepts covered in this book and on the exams may seem pretty elementary to you. But if you haven't worked with them for awhile, a brush-up is probably a good idea. Also, standardized tests cover math in ways that are a bit quirky and that may catch you off guard if you're not ready. Finally, keep in mind that standardized testing is a "game of inches" in which just one or two additional correct responses can significantly enhance your scaled score and thus your chances of gaining admission to your first-choice school. So even if this book helps you get two or three questions correct that you otherwise would have missed, it's well worth the effort.

Q: *Isn't the best way to prepare for the math portion of my exam simply to take as many practice tests as possible?*

A: Definitely not. Sure, taking practice tests will help you to become familiar and comfortable with your exam's format and time restrictions, as well as reinforcing math concepts you're reviewing for your exam. But it won't necessarily help you improve your score. In fact, you're likely to just keep repeating the same mistakes over and over.

Reading explanations for the questions that stump you is helpful, of course, but it's not enough. You may say to yourself after reading the explanation: "Okay, I've got it now; I won't miss that kind of question again." Wrong! You probably won't encounter just "that kind of question" on your actual exam. Even subtle variations in problems may throw you for a loop unless you've truly mastered the concept underlying the problems. So the key is not how *much* you prepare, but how *smart* you prepare.

Q: *Are there certain basic math skills I should have before I start working though this book?*

A: Yes. We won't be going all the way back to basics here. Armed with paper and pencil, you should already be able to perform these arithmetical operations:

- adding and subtracting columns of numbers
- performing long division
- multiplying large numbers together

If you need help with these basic skills, put this book down and find a good arithmetic workbook. What you will learn in this book—specifically, in Chapter 9—is how to do arithmetic *quicker and more accurately* than the old-fashioned ways you learned in grade school. In other words, you'll be learning the test-taking skills that will set you apart from the crowd on your exam.

Q: *Will I have to memorize math "formulas" and computational tables to score high on my exam?*

A: You will have to memorize certain algebraic and geometric formulas to perform well on your exam. All the ones you need to know are in this book. These formulas are covered extensively on the exams. (Any study guide that doesn't focus on the formulas belongs in the trash can!)

Now for some good news: you won't have to memorize extensive computational tables for your exam. Just make sure you know the brief tables of "easier" numbers included throughout this book, and you'll be all set. The test-makers are trying to gauge your understanding of underlying concepts, *not* your ability to "crunch" large and unwieldy numbers. So they design test questions using easier numbers and numbers that you can round off to determine the correct response.

Q: *Will I be learning tricks, secrets, and shortcuts in this book to help me "crack" the test or to beat the test-maker at its "own game"?*

A: No, and it would be futile for you to try to do so. Be assured: The test-makers do not design their tests in a manner that leaves them vulnerable to safe-cracking artists. Sure, there are techniques and strategies to save time and improve accuracy on these tests. (In Chapter 9, you'll learn such techniques.) But the hard truth is that there are no shortcuts to understanding the underlying mathematical concepts being tested.

Q: *Besides working through this book, what else should I do to prepare for the math portion of my exam?*

A: By all means, make sure you're familiar with the specific format of the math section(s) on your exam. Read carefully the descriptions and instructions provided by the testing service in its official printed publications or at its Web site. Take at least two or three practice tests that accurately simulate your particular exam. (Publications containing simulated tests are readily available at general bookstores.) Be sure to take practice tests under simulated testing conditions. If you're planning to take a computerized version of the GRE or GMAT, it's a good idea to use appropriate software for simulated testing.

How to Get the Most Out of This Book

You'll benefit most from this book by working through each chapter in sequence, without skipping around. Why? Because more advanced math concepts build upon more basic ones. Accordingly, you should postpone your study of algebra and geometry (Chapters 10–15) until you've mastered more fundamental concepts (Chapters 3–9). Here's a more detailed look at what to expect.

Chapter 2: Test Your Math Smarts. Here you'll find a 40-question multiple-choice quiz that covers all of the broad areas of your exam. It's a great place to assess your strengths and weaknesses so that you know where to concentrate your attention later in the book.

Chapters 3–9. These seven chapters are all about numbers and arithmetic. We start with numbers because numbers are the backbone of all areas of mathematics. If you can't quickly and accurately add, subtract, multiply, divide, and work with exponents and roots, then forget about more advanced fare such as algebra and geometry. That would be like trying to play a one of Beethoven's piano concertos before you've learned the eight-note major scale. Pay close attention to these earlier chapters, since any misunderstanding or confusion about basic concepts can throw your math score into a tailspin.

Chapters 10–12. These three chapters focus on algebra. Chapter 10 lays the foundation, while 11 and 12 apply the basics to statistical problems and to those "real-life" word problems that can be so troublesome.

Chapters 13–15. These three chapters focus on geometry. Don't skip past algebra to these chapters; you'll need solid algebra skills before tackling geometry. The various geometric figures examined will grow increasing complex as you move along, from intersecting lines (Chapter 13) to those pesky 4-sided pyramids (Chapter 15).

The "Quiz Time" Quizzes. At the conclusion of each chapter (except for Chapter 8), you'll find a 10-question quiz to help you review the concepts covered in the chapter. Every quiz includes 5 "easier" questions and 5 "more challenging" questions.

Note: The quizzes are not designed to emulate the format of any particular exam. You'll find a variety of formats here. Some questions are multiple-choice, others require your own response. Some are multi-part, while others are not. Be sure to read the explanations for questions that stump you.

A Little Math Before Moving On

Let's not leave this introductory chapter without a small taste of math. As noted earlier, to save time on your exam you should be able to perform arithmetical operations on common numbers quickly in your head. Start by making the "times" table second nature to you, *not* just 1 though 10, but through 12. You'll save precious time on your exam if you can rattle off these common multiples in the blink of an eye.

Memorize this multiplication table:

	1	2	3	4	5	6	7	8	9	10	11	12
1	1	2	3	4	5	6	7	8	9	10	11	12
2	2	4	6	8	10	12	14	16	18	20	22	24
3	3	6	9	12	15	18	21	24	27	30	33	36
4	4	8	12	16	20	24	28	32	36	40	44	48
5	5	10	15	20	25	30	35	40	45	50	55	60
6	6	12	18	24	30	36	42	48	54	60	66	72
7	7	14	21	28	35	42	49	56	63	70	77	84
8	8	16	24	32	40	48	56	64	72	80	88	96
9	9	18	27	36	45	54	63	72	81	90	99	108
10	10	20	30	40	50	60	70	80	90	100	110	120
11	11	22	33	44	55	66	77	88	99	110	121	132
12	12	24	36	48	60	72	84	96	108	120	132	144

2

Test Your
Math Smarts
(with this 40-Question Quiz)

L et's find out right away if you have what it takes to be a smart test-taker when it comes to math. The 40 multiple-choice questions here cover the math topics in this book and on most standardized math exams.

Keep in mind: This quiz is designed to introduce you to the topics covered on standardized math exams and provide immediate feedback about your strengths and weaknesses. It is *not* designed to simulate the *format* of the actual exams (such as the SAT, GRE, ACT, and GMAT), which vary in number of questions as well as question format.

Your performance on this warm-up is your cue for self-study. After attempting all 40 questions, read the explanations (beginning on page 15). Don't worry too much about your "score." If you don't understand a certain question or explanation, that's your cue to focus your attention on that particular topic as you work your way through Chapters 3–15.

Directions: Make sure that you have 60 minutes of uninterrupted time, pick up a pencil, and put on your thinking cap. Circle the correct answer choice for each question. Go for it!

1.   Which of the following fractions is the smallest?

   (A) $\frac{3}{4}$ (B) $\frac{5}{6}$ (C) $\frac{7}{8}$ (D) $\frac{19}{24}$ (E) $\frac{13}{15}$

2.   If $x + 3$ is a multiple of 3, which of the following is not a multiple of 3?

   (A) x (B) $x + 6$ (C) $6x + 18$ (D) $2x + 6$ (E) $3x + 5$

3.   If $x < 0$ and $0 < y < 1$, which of the following has the largest value?

   (A) $x + y$ (B) $y - x$ (C) $\frac{x}{y}$ (D) xy (E) $x^3 - y$

4.   If $m = n$ and $p < q$, then

   (A) $m - p < n - q$ (B) $p - m > q - n$ (C) $m - p > n - q$

   (D) $mp > nq$ (E) $m + q < n + p$

5.   If the legislature passes a particular bill by a ratio of 5 to 4, and if 900 legislators voted in favor of the bill, how many voted against it?

   (A) 400   (B) 500   (C) 720   (D) 760   (E) 800

6.   If a certain gas tank is $\frac{1}{4}$ full, and after adding G gallons of gas the tank is $\frac{7}{8}$ full, what is the capacity of the tank?

   (A) $\frac{5G}{8}$ (B) $\frac{8G}{5}$ (C) $\frac{8G}{7}$ (D) $\frac{7G}{8}$ (E) $4G$

7. M college students agree to rent an apartment for D dollars per month, sharing the rent equally. If the rent is increased by $100, what amount must each student contribute?

   (A) $\frac{D}{M}$ (B) $\frac{D}{M} + 100$ (C) $\frac{D+100}{M}$ (D) $\frac{M}{D} + 100$ (E) $\frac{M+100}{D}$

8. A gear having 20 teeth turns at 30 revolutions per minute and is meshed with another gear having 25 teeth. At how many revolutions per minute is the second gear turning?

(A) $22\frac{1}{2}$ (B) 24 (C) 30 (D) 35 (E) $37\frac{1}{2}$

9. A photographic negative measures $1\frac{7}{8}$ inches by $2\frac{1}{4}$ inches. If the longer side of the printed picture is to be 4 inches, how many inches long will the shorter side of the printed picture be?

(A) $2\frac{3}{8}$ (B) $2\frac{1}{2}$ (C) 3 (D) $3\frac{1}{8}$ (E) $3\frac{3}{8}$

10. M is P% of what number?

(A) $\frac{MP}{100}$ (B) $\frac{100P}{M}$ (C) $\frac{M}{100P}$ (D) $\frac{P}{100M}$ (E) $\frac{100M}{P}$

11. How many fifths are in 280%?

(A) 1.4 (B) 2.8 (C) 14 (D) 28 (E) 56

12. If the value of XYZ Company stock drops from $25 per share to $21 per share, what is the percent of decrease?

(A) 4 (B) 8 (C) 12 (D) 16 (E) 20

13. Bobby sent $27 to the newspaper dealer for whom he delivers papers, after deducting a 10% commission. How many papers did he deliver if papers sell for 20 cents each?

(A) 135 (B) 150 (C) 160 (D) 540 (E) 600

14. What is the average (arithmetic mean) of $\sqrt{.49}$, $\frac{3}{4}$, and 80%?

(A) .073 (B) .075 (C) .72 (D) .75 (E) .78

15. The average of seven numbers is 84. Six of the numbers are: 86, 82, 90, 92, 80, and 81. What is the seventh number?

(A) 79 (B) 85 (C) 81 (D) 77 (E) 76

16. At ABC Corporation, five executives earn \$150,000 each per year, three executives earn \$170,000 each per year, and one executive earns \$180,000 per year. What is the average salary of these executives?

(A) \$156,250   (B) \$160,000   (C) \$164,480

(D) \$166,670   (E) \$170,000

17. If $4(x - r) = 2x + 10r$, then $x =$

(A) r (B) $2\frac{1}{3}r$ (C) $3r$ (D) $5.5r$ (E) $7r$

18. If $.2y = 2.2 - .6x$ and $.5x = .2y + 1.1$, then $x =$

(A) 1   (B) 3   (C) 10   (D) 11   (E) 30

19. If $19x - 48 = x^2$, then the possible values of x include

(A) 8 and 6   (B) 24 and 2   (C) −16 and −3

(D) 12 and 4   (E) 16 and 3

20. If $\sqrt{x^2 + 3} = x + 1$, then $x =$

(A) −3   (B) −1   (C) 0   (D) 1   (E) 2

21. If $25x^2 = 4$, then the possible values of x include

(A) $\frac{4}{25}$ and $-\frac{4}{25}$ (B) $\frac{2}{5}$ and $-\frac{2}{5}$ (C) 2 and 5

(D) $\frac{2}{5}$ only   (E) $-\frac{2}{5}$ only

22. $\frac{1}{2}\sqrt{180} + \frac{1}{3}\sqrt{45} - \frac{2}{5}\sqrt{20}$ is equivalent to which of the following?

(A) $3\sqrt{10} + \sqrt{15} - 2\sqrt{2}$ (B) $\frac{16}{5}\sqrt{5}$ (C) $\sqrt{97}$ (D) $\frac{24}{5}\sqrt{5}$ (E) 30

23. $\sqrt{\dfrac{y^2}{2} - \dfrac{y^2}{18}} =$

(A) 0   (B) $\dfrac{2y}{3}$ (C) $\dfrac{10y}{3}$ (D) $\dfrac{y\sqrt{3}}{6}$ (E) $\dfrac{y\sqrt{5}}{3}$

24. $\dfrac{2 + \dfrac{1}{t}}{\dfrac{2}{t^2}} =$

(A) $t^2 + t$ (B) t^3 (C) $\dfrac{2t+1}{2}$ (D) $t+1$ (E) $\dfrac{t(2t+1)}{2}$

25. In a purse containing only nickels and dimes, the ratio of nickels to dimes is 3:4. If there are 28 coins altogether, what is the value of the dimes?

(A) $1.20   (B) $1.40   (C) $1.60   (D) $1.70   (E) $2.10

26. If n is the first of two consecutive odd integers, which equation could be used to find these two integers if the difference of their squares is 120?

(A) $(n+1)^2 - n^2 = 120$ (B) $n^2 - (n+2)^2 = 120$ (C) $[(n+2) - n]^2 = 120$

(D) $(n+2)^2 - n^2 = 120$ (E) $n^2 - (n+1)^2 = 120$

27. Lyle is 23 years old and Melanie is 15 years old. How many years ago was Lyle twice as old as Melanie?

(A) 5   (B) 7   (C) 8   (D) 9   (E) 16

28. A portion of $7,200 is invested at a 4% annual return, while the remainder is invested at 5% annual return. If the annual income from both portions is the same, what is the total income from the two investments?

(A) $160   (B) $320   (C) $400   (D) $720   (E) $1,600

29. The denominator of a fraction is twice as large as the numerator. If 4 is added to both the numerator and denominator, the value of the fraction is $\frac{5}{8}$. The denominator of the original fraction is

(A) 6   (B) 10   (C) 12   (D) 14   (E) 16

30. A container holds 10 liters of a solution which is 20% acid. If 6 liters of pure acid are added to the container, what percent of the resulting mixture is acid?

(A) 5   (B) 10   (C) 20   (D) $33\frac{1}{3}$ (E) 50

31. At 10 a.m. two trains started traveling toward each other from stations 287 miles apart. They passed each other at 1:30 p.m. the same day. If the average speed of the faster train exceeded the average speed of the slower train by 6 miles per hour, which of the following represents the speed of the faster train, in miles per hour?

(A) 38   (B) 40   (C) 44   (D) 48   (E) 50

32. A swimming pool can be filled through an inlet pipe in 3 hours. It can be drained by a drainpipe in 6 hours. If both pipes are fully open at the same time, and if the pool is empty, in how many hours will it be filled?

(A) 4   (B) 4.5   (C) 5   (D) 5.5   (E) 6

33. A farmer uses 140 feet of fencing to enclose a rectangular field. If the length of the field is $33\frac{1}{3}$% greater than its width, what is the diagonal distance from one corner of the enclosure to the opposite corner?

(A) 10   (B) 20   (C) 50   (D) 70   (E) 100

34. In parallelogram *ABCD*, angle *B* is 5 times as large as angle *C*. What is the measure, in degrees, of angle *C*?

(A) 30   (B) 45   (C) 80   (D) 120   (E) 150

35.   In the diagram below, if AB is parallel to CD, what is the degree measure of $\angle BED$? (Note: the figure is not drawn to scale.)

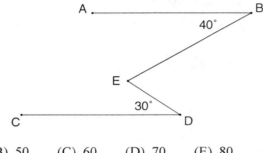

(A) 40   (B) 50   (C) 60   (D) 70   (E) 80

36.   In a circle whose center is O, arc AB measures $85°$. What is the degree measure of $\angle ABO$?

(A) 45   (B) 47.5   (C) 85   (D) 95   (E) 137.5

37. AB is the diameter of a circle whose center is O. If the coordinates of O are $(2,1)$ and the coordinates of B are $(4,6)$, what are the coordinates of A?

(A) $(3, 3\frac{1}{2})$ (B) $(1, 2\frac{1}{2})$ (C) $(0, -4)$ (D) $(2\frac{1}{2}, 1)$ (E) $(-1, -2\frac{1}{2})$

38.   In the figure below, what is the length of DB?

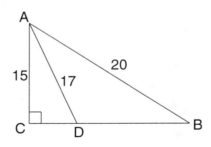

(A) $5\sqrt{21} - 8$ (B) 12   (C) $8\sqrt{3}$ (D) $5\sqrt{7} - 8$ (E) $18 - 2\sqrt{6}$

39. A rectangular tank 10" by 8" by 4" is filled with water. If all of the water is to be transferred to cube-shaped tanks, each one 3 inches on a side, how many of these smaller tanks are needed?

(A) 9   (B) 12   (C) 16   (D) 39   (E) 21

40. In the figure below, if $DC = 12$, what is the area of $ABCD$?

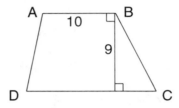

(A) 99   (B) 108   (C) 112   (D) 120   (E) $50\sqrt{3}$

STOP.

Quick Answer Key

1. A	11. C	21. B	31. C
2. E	12. D	22. B	32. E
3. B	13. B	23. B	33. C
4. C	14. D	24. E	34. A
5. C	15. D	25. C	35. D
6. B	16. B	26. D	36. B
7. C	17. E	27. B	37. C
8. B	18. B	28. B	38. D
9. C	19. E	29. C	39. B
10. E	20. D	30. E	40. A

Explanations

1. (A) To compare (A), (B), (C), and (D), use a common denominator of 24:

$$\frac{3}{4} = \frac{18}{24} \qquad \frac{5}{6} = \frac{20}{24} \qquad \frac{7}{8} = \frac{21}{24} \qquad \frac{19}{24}$$

Of these four numbers, $\frac{3}{4}$ is the smallest. To compare $\frac{3}{4}$ with $\frac{13}{15}$, compare "cross products." Since $(3)(15) < (4)(13)$, $\frac{3}{4} < \frac{13}{15}$.

2. (E) $3x$ is a multiple of 3; thus, adding five to that number will yield a number that is not a multiple of 3.

3. (B) Answer choices (A), (C), (D), and (E) are each negative in value, while (B) must be positive.

4. (C) In answer choice (C), unequal quantities are subtracted from equal quantities. The differences are unequal, but the inequality is reversed since unequal numbers are being subtracted from rather than added to the equal numbers.

5. (C) $\frac{5}{9}$ of the legislators voted for the bill. Determine the total number of legislators by setting up and solving the following equation, then subtract 900 from the total:

$$900 = \tfrac{5}{9}x$$
$$8100 = 5x$$
$$1620 = x$$
$$1620 - 900 = 720$$

6. (B) The G gallons that are added fill $\frac{7}{8} - \frac{1}{4}$ or $\frac{5}{8}$ of the tank.

$$\frac{5}{8}x = G$$
$$x = \frac{8G}{5}$$

7. (C) The total rent is $D + 100$, which must be divided by the number of students.

8. (B) The number of teeth multiplied by speed remains constant:

$$(20)(30) = 25x$$
$$600 = 25x$$
$$x = 24$$

9. (C) Equate the proportions of the negative with those of the printed picture:

$$\frac{2\frac{1}{2}}{4} = \frac{1\frac{7}{8}}{x}$$
$$\frac{\frac{5}{2}}{4} = \frac{\frac{15}{8}}{x}$$
$$\frac{5}{2}x = \frac{15}{2}$$
$$5x = 15$$
$$x = 3$$

10. (E) Convert the question into an algebraic equation, and solve for x:

$$M = \frac{P}{100} \cdot x$$
$$100 = Px$$
$$\frac{100M}{P} = x$$

11. (C) $280\% = \frac{280}{100} = \frac{28}{10} = \frac{14}{5}$

12. (D) The amount of the decrease is \$4. The percent of the decrease is $\frac{4}{25}$, or $\frac{16}{100}$, or 16%.

13. (B) \$27 is 90% of what he collected. Express this as an equation:

$$27 = .90x$$
$$270 = 9x$$
$$x = \$30$$

If each paper sells for 20 cents, he sold $\frac{30.00}{.20}$ or 150 papers.

14. (D) $\sqrt{.49} = .7$, $\frac{3}{4} = .75$, and $80\% = .8$. Their sum is 2.25. The average is $2.25 \div 3$, or .75.

15. (D) You could solve the problem algebraically by using the arithmetic mean formula (x is the seventh number):

$$84 = \frac{86 + 82 + 90 + 92 + 80 + 81 + x}{7}$$

There's a quicker way, however. 86 is 2 above the 84 average, and 82 is two below. These two numbers "cancel" each other. 90 is 6 above and 92 is 8 above the average (a total of 14 above), while 80 is 3 below and 81 is 3 below the average (a total of 7 below). Thus the six terms average out to 7 above the average of 84. Accordingly, the seventh number is 7 below the average of 84, or 77.

16. (B) Assign a "weight" to each of the three salary figures, then determine the weighted average of the nine salaries:

$$5(150,000) = 750,000$$

$$3(170,000) = 510,000$$

$$1(180,000) = 180,000$$

$$750,000 + 510,000 + 180,000 = 1,440,000$$

$$\frac{1,440,000}{9} = 160,000$$

17. (E)

$$4x - 4r = 2x + 10r$$
$$2x = 14r$$
$$x = 7r$$

18. (B) Because the y-terms are the same $(.2y)$, the quickest way to solve for x here is with the addition-subtraction method:

$$.2y + .6x = 2.2$$
$$\underline{-.2y + .5x = 1.1}$$
$$1.1x = 3.3$$
$$x = 3$$

19. (E) Put this quadratic equation into the form: $ax^2 + bx + c = 0$, then factor the trinomial into two binomials:

$$x^2 - 19x + 48 = 0$$
$$(x - 16)(x - 3) = 0$$
$$x - 16 = 0 \text{ or } x - 3 = 0$$
$$x = 16 \text{ or } 3$$

20. (D) Square both sides of the equation, then solve for x:

$$\left(\sqrt{x^2 + 3}\right)^2 = (x + 1)^2$$
$$x^2 + 3 = (x + 1)(x + 1)$$
$$x^2 + 3 = x^2 + 2x + 1$$
$$2x = 2$$
$$x = 1$$

21. (B) You can solve this problem quickly by recognizing that the quadratic equation is of the form: $x^2 - y^2 = 0$. Since $x^2 - y^2 = (x + y)(x - y)$:

$$(5x)^2 - 2^2 = 0$$
$$(5x + 2)(5x - 2) = 0$$
$$5x = 2 \text{ or } -2$$
$$x = \tfrac{2}{5} \text{ or } -\tfrac{2}{5}$$

22. (B) In each of the three terms, the radical value can be factored:

$$\tfrac{1}{2}\sqrt{180} = \tfrac{1}{2}\sqrt{36 \cdot 5} = 3\sqrt{5}$$
$$\tfrac{1}{3}\sqrt{45} = \tfrac{1}{3}\sqrt{9 \cdot 5} = \sqrt{5}$$
$$\tfrac{2}{5}\sqrt{20} = \tfrac{2}{5}\sqrt{4 \cdot 5} = \tfrac{4}{5}\sqrt{5}$$

Combining these three simplified terms:
$$3\sqrt{5} + \sqrt{5} - \tfrac{4}{5}\sqrt{5} = 3\tfrac{1}{5}\sqrt{5} \text{ or } \tfrac{16}{5}\sqrt{5}$$

23. (B) Combine the terms under the radical into one fraction, then factor out "squares" from both numerator and denominator:

$$\sqrt{\frac{y^2}{2} - \frac{y^2}{18}} = \sqrt{\frac{9y^2 - y^2}{18}} = \sqrt{\frac{8y^2}{18}} = \sqrt{\frac{4y^2}{9}} = \frac{\sqrt{4y^2}}{\sqrt{9}} = \frac{2y}{3}$$

24. (E) To eliminate the complex fraction, multiply the two terms in the numerator as well as the denominator fraction by t^2. Then factor out a t from each term in the numerator:

$$\frac{t^2(2+\frac{1}{t})}{2} = \frac{2t^2 + \frac{t^2}{t}}{2} = \frac{2t^2 + t}{2} = \frac{t(2t+1)}{2}$$

25.   (C) Let $3x$ equal the number of nickels, and let $4x$ equal the number of dimes.

$$3x + 4x = 28$$
$$7x = 28$$
$$x = 4$$

The number of dimes is $4(4) = 16$. Thus, the total value of the dimes is \$1.60.

26.   (D) The other integer is $n + 2$. Since the difference between n and $n + 2$ is positive, the larger number, which is $n + 2$, must come first in the equation.

27.   (B) You can solve the problem algebraically as follows:

$$23 - x = 2(15 - x)$$
$$23 - x = 30 - 2x$$
$$x = 7$$

An alternative method is to subtract the number in each answer choice from both Lyle's age and Melanie's age.

28.   (B) Letting x equal the amount invested at 4%, express the amount invested at 5% as $7200 - x$. The return on these amounts is equal:

$$.04x = .05(7200 - x)$$

Multiply by 100 to eliminate decimals:

$$4x = 5(7,200 - x)$$
$$4x = 36,000 - 5x$$
$$9x = 36,000$$
$$x = 4,000$$

The income is $.04(4,000) + .05(3,200) = \$160 + \$160$, or \$320.

29.   (C) Represent the original fraction by $\frac{x}{2x}$:

$$\frac{x+4}{2x+4} = \frac{5}{8}$$

Cross multiply, and solve for x:

$$8x + 32 = 10x + 20$$
$$12 = 2x$$
$$x = 6$$

The original denominator is $2x$, or 12.

30. (E) The original amount of acid is $(10)(20\%) = 2$ liters. After adding 6 liters of pure acid, the amount of acid increases to 8 liters, while the amount of total solution increases from 10 to 16 liters. The new solution is $\frac{8}{16}$ or 50% acid.

31. (C) The trains each traveled from 10 a.m. to 1:30 p.m., which is 3.5 hours. Let x equal the speed of the slower train, and let $x + 6$ equal the speed of the faster train:

$$3.5x + 3.5(x + 6) = 287$$

Multiply by 10 to eliminate decimals, then solve for x:

$$35x + 35(x + 6) = 2870$$
$$35x + 35x + 210 = 2870$$
$$70x = 2{,}660$$
$$x = 38$$

The speed of the faster train was $x + 6$ or 44 m.p.h.

32. (E) Letting x equal the number of hours, subtract the drainpipe's rate from the inlet pipe's rate (subtract because the drainpipe works against the inlet pipe), using the "work" formula:

$$\frac{x}{3} - \frac{x}{6} = 1$$

Multiply both sides by 6, then solve for x:

$$2x - x = 6$$
$$x = 6$$

33. (C) The length is $\frac{4}{3}$ the width, so the ratio of length to width is 4:3. Let $3x$ equal the width, and let $4x$ equal the length. Given a perimeter of 140, you can determine the length and width as follows:

$$2(3x + 4x) = 140$$
$$7x = 70$$
$$x = 10$$

length = 40, width = 30

The diagonal distance from one corner to the other is the hypotenuse of a 3:4:5 right triangle. Thus, that distance is 50.

34. (A) Let x equal the measure of angle C, and let $5x$ equal the measure of angle B. The sum of the angles of a parallelogram is 360.

$$5x + 5x + x + x = 360$$
$$12x = 360$$
$$x = 30$$

35. (D) Extend *BE* to *F* (as in the diagram below). $\angle EFD = \angle ABE = 40°$. $\angle FED$ must equal 110° because there are 180° in a triangle. Since $\angle BED$ is the supplement of $\angle FED$, $\angle BED = 70°$.

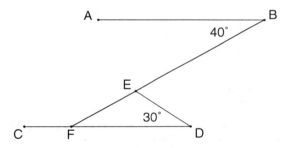

36. (B) Angle O is the central angle equal to its arc, 85°. Since the sum of all three angles is 180, the sum of the remaining two angles is 95. The triangle is isosceles because the legs are both radii. Thus, the two angles are equal in size; each one is 47.5 degrees.

37. (C) O is the midpoint of *AB*. Thus, the run (horizontal distance) and rise (vertical distance) from B to O are the same as from O to A (averaging the x-coordinates and averaging the y-coordinates). Those distances are 2 and 5, respectively. From O (2,1), move two units to the left and 5 units down, to (0,–4).

38. (D) To determine DB, subtract CD from CB. Thus, we need to find those two lengths first. $\triangle ACD$ is a right triangle with sides 8, 15, and 17 (one of the Pythagorean triplets). Thus, $CD = 8$. CB is one of the legs of $\triangle ABC$. Determine CB by applying the Theorem:

$$15^2 + (CB)^2 = 20^2$$
$$225 + (CB)^2 = 400$$
$$(CB)^2 = 175$$
$$CB = \sqrt{25 \cdot 7} = 5\sqrt{7}$$

Accordingly, $CD = 5\sqrt{7} - 8$.

39. (B) The volume of the large tank is $10 \cdot 8 \cdot 4$, or 320, cubic inches. The volume of each small cube is 3^3, or 27, cubic inches. 12 cube-shaped tanks are needed. ($27 \cdot 12 = 324$). The last cube-shaped tank will have 4 cubic inches of unused capacity.

40. (A) Because of the two right angles indicated in the figure, $AB \parallel DC$, and $ABCD$ is a trapezoid. The area of a trapezoid $= \frac{1}{2}h(b_1 + b_2)$, where h is the height and each b is a parallel base (side):

$$A = \frac{1}{2}(9)(10 + 12) = 99$$

3

Basic Stuff
about Numbers
and Arithmetic

D on't get the idea from the title of this chapter that you can just skip over it to Chapter 4. Sure, you'll probably find *some* of the "basic stuff" here pretty easy. But you'll also learn to look at numbers and arithmetic in a new and more intuitive way—a way that helps you to handle numbers and arithmetic problems more effectively. Besides, if you misunderstand the "basic" concepts covered here, you'll be in big trouble when we get to more advanced concepts in subsequent chapters. **In this chapter**, you'll be exploring these concepts:

- real numbers, absolute value, and number "signs"

- factors, multiples, and prime numbers

- rules for combining numbers (and other terms) using the four basic arithmetical operations

- shortcuts for combining numbers

Don't forget to take the **10-question quiz** (starting on page 41) after reading the chapter to review the concepts you just learned.

Real Numbers, Absolute Value, and Number "Signs"

First things first: Let's make sure we're clear about what we mean when we say "numbers." There are many different kinds of numbers and many different ways of representing them. The numbers that appear on standardized math exams are limited to **real numbers**. A real number is any number on the real number line:

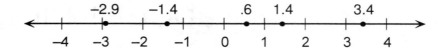

On the real number line, remember that "left is less." In other words, a real number is *less* than another real number if it appears to the *left* of the other number on the number line. For example, if $x < y < z$, then x is to the left of both y and z on the number line, while y is to the left of z on the number line. In other words, y is between x and z.

On the real number line, the number "zero" (0) is especially important; it marks the dividing line between *negative* numbers (numbers "less than zero"—left of zero on the number line) and *positive* numbers (numbers "greater than zero"—right of zero on the number line). On the number line, zero is also referred to as the *origin*. All real numbers except zero are either positive or negative. Any negative number is less than any positive number:

$$\{\text{negative numbers}\} \ < \ \text{zero} \ < \ \{\text{positive numbers}\}$$

Absolute Value

The absolute value of a real number refers to the number's *distance from zero* (the origin) on the number line. The symbol for absolute value is a pair of vertical lines: | |. Although any negative number is less than any positive number, the absolute value of a negative number may be less than, equal to, or greater than a positive number. Here is one instance of each:

$$|-7| = 7 < 8 \qquad |-7| = 7 \qquad |-7| = 7 > 6$$

Imaginary Numbers

Okay, now we know what real numbers are, but what kinds of numbers are "unreal"? In basic math, you'll run across only two examples of "unreal" or **imaginary** numbers (as mathematicians call them):

$$\frac{0}{0} = \text{indeterminate (an imaginary number)}$$

$$\frac{n}{0} = \text{undefined, where } n \neq 0 \text{ (an imaginary number)}$$

Why don't these fractions equal zero, you ask? Well, think about it. What numbers multiplied by zero (0) would result in a product of zero (0)? *Any* real number would do, wouldn't it? In other words, there are an infinite number of answers. What numbers divided by zero would result in a non-zero product? *No* real number would fit the bill, would it? So our real number system has no way of accommodating fractions with a denominator of zero.

So what do we do about these odd ducks? Relax; you won't need to deal with them in SAT, ACT, GRE, or GMAT math sections. These imaginary numbers are "written out" of the exams, in two ways:

1. The general instructions will state that all numbers used are real numbers.

2. Any specific problem that might involve a zero denominator will explicitly rule out this possibility by restricting the denominator's value. A question might state, for example:

$$\text{Q: If } \frac{y}{x} = z \text{ and } xyz \neq 0 \text{ ...}$$

In this example, $\frac{y}{x} = z$ can also be expressed as $x = \frac{y}{z}$ or $1 = \frac{y}{xz}$. Thus, if either x, y, or z were to equal zero, the problem would involve an imaginary number. In stipulating that $xyz \neq 0$, the problem eliminates the possibility that any one of the three variables equals zero (zero × any real number = zero).

Number Signs and the Four Basic Operations

A number's "sign" designates whether the number is positive or negative. What happens to a number's sign when you combine it with another number, either by addition, subtraction, multiplication, or division? Here are the possibilities:

[Note: (+) signifies any positive number, and (–) signifies any negative number]

multiplication: (examples)

$(+) \cdot (+) = (+)$ $4 \cdot 2 = 8$

$(+) \cdot (–) = (–)$ $4 \cdot (–2) = –8$

division:

$(+) \div (+) = (+)$ $4 \div 2 = 2$

$(+) \div (–) = (–)$ $4 \div (–2) = –2$

$(–) \div (+) = (–)$ $–4 \div 2 = –2$

addition:

$(+) + (+) = (+)$ $4 + 2 = 6$

$(+) + (–) = ?$ $4 + (–2) = 2$, but $2 + (–4) = –2$

subtraction:

$(+) – (–) = (+)$ $4 – (–2) = 4 + 2 = 6$

$(–) – (+) = (–)$ $–4 – 2 = –6$

$(–) – (–) = ?$ $–4 – (–2) = –2$, but $–2 – (–4) = 2$

You can visualize the rules for addition and subtraction on the number line itself, moving either to the left or right as you add or subtract numbers. You can't do this with multiplication and division, however. Let's look again at multiplication and division, taking the rules for signs a step further. Because multiplying (or dividing) two negative terms always results in a positive sign, combining two or more *pairs* of negative terms (in other words, any *even* number of negative terms) by multiplication and division must also result in a positive number. Here are some examples:

even number of negative terms:	odd number of negative terms:
$(-) \cdot (-) = (+)$	$(-) \div (-) \div (-) = (-)$
$(-) \cdot (-) \div (-) \cdot (-) = (+)$	$(-) \div (-) \div (-) \cdot (-) \div (-) = (-)$
$(-) \div (-) \div (-) \cdot (-) \cdot (-) \cdot (-) = (+)$	

This rule doesn't work for addition and subtraction, of course. For those operations, the resulting sign depends entirely on the size of the numbers.

Integers

Real numbers include **integers** and **non-integers**. Integers are the "counting" numbers on the number line:

$$\{ \dots -3, \ -2, \ -1, \ 0, \ 1, \ 2, \ 3, \ \dots \}$$

Non-integers include all other real numbers, such as fractions that cannot be "reduced" (simplified) to integers as well as decimal numbers (numbers with non-zero digits to the right of a decimal point); for example:

$$-13\tfrac{3}{5} \qquad\qquad 78.3$$

Except for the number zero (0), every integer is either positive or negative and either even or odd:

negative integers:	$\{\dots-3, -2, -1\}$
positive integers:	$\{1, 2, 3,\dots\}$
non-negative integers:	$\{0, 1, 2, 3,\dots\}$
even integers:	$\{-4, -2, 2, 4\dots\}$
odd integers:	$\{-3, -1, 1, 3,\dots\}$

Remember: Zero is the only integer that is neither even nor odd and that is neither positive nor negative.

Patterns to Look for When Combining Integers

What happens to integers when you combine them by adding, subtracting, multiplying, or dividing? In addition to the rules for signs you just learned (which apply to integers and non-integers alike), here are some useful observations that apply *only to integers*:

addition and subtraction:

- integer $\pm$ integer = integer
- even $\pm$ even = even (or zero, if the two integers are the same)
- even $\pm$ odd = odd
- odd $\pm$ odd = even (or zero, if the two integers are the same)

multiplication and division:

- (integer) $\times$ (integer) = (integer)
- (integer) $\div$ (non-zero integer) = (integer), but only if the numerator is divisible by the denominator (if the result is a quotient with no remainder)
- (odd integer) $\times$ (odd integer) = (odd integer)
- (even integer) $\times$ (non-zero integer) = (even integer)
- (even integer) $\div$ (2) = (integer)
- (odd integer) $\div$ (2) = (non-integer)

Special Characteristics of 1 and 0 (Zero)

Here are three unique characteristics of the integer 1 that you should keep in mind:

$$n \times 1 = n \qquad \frac{n}{n} = 1 \ (n \neq 0) \qquad \frac{n}{1} = n$$

Here are three unique characteristics of the integer 0 (zero):

$$n \pm 0 = n \qquad n \times 0 = 0 \qquad \frac{0}{n} = 0$$

Here are two more characteristics of the integer 0 (zero) that you should distinguish from the preceding ones:

$$\frac{n}{0} = \text{undefined} \qquad\qquad \frac{0}{0} = \text{indeterminate}$$

As noted on page 27, you will not encounter an undefined or indeterminate number on most standardized exams; questions will explicitly restrict denominators to non-zero values to avoid these "imaginary" numbers.

Factors, Multiples, and Prime Numbers

The best way to understand the meaning of the terms *factor*, *multiple*, and *divisibility* (as they're used in math) is to look at a simple example. Consider these two equations:

$$2 \cdot 3 = 6 \qquad\qquad 6 \div 3 = 2$$

The integer 2 is said to be a "factor" of the integer 6 because you can multiply 2 by an *integer* to obtain 6 (or because you can divide 6 by an *integer* to obtain 2). For the same reason, 6 is said to be "a *multiple* of" or "*divisible* by" 2. Similarly:

3 is a *factor* of 6

6 is a *multiple* of 3

6 is *divisible* by 3

Just for the record, let's state all of this a bit more formally. If you divide an integer *m* by a second integer *f*, and if the result is a third integer (with no remainder), then *m* is a multiple of (is divisible by) *f* and, conversely, *f* is a factor of *m*.

The concept of divisibility is essential to basic mathematics. Whenever you multiply one integer by another, you are combining two factors to obtain a *common multiple*:

$$3 \cdot 13 = 39 \qquad\qquad \text{3 and 13 are factors of 39}$$

Conversely, whenever you divide one integer by another, you are determining whether the second number is a factor of the first number. If the result is an integer with no remainder, then the quotient and the second number are both factors of the first number.

$$48 \div 8 = 6$$ 6 and 8 are factors of 48

$$48 \div 5 = 9\tfrac{3}{5}$$ 5 is not a factor of 48 because $9\tfrac{3}{5}$ is not an integer.

Whenever you "reduce" a fraction (replace both the numerator and denominator with smaller numbers), you are "canceling" or "factoring out" a common factor in the numerator and denominator:

$$\frac{14}{63} = \frac{(7)(2)}{(7)(9)} = \frac{2}{9}$$ 7 is a factor of both 14 and 49, so it can be "factored out."

Here are some important points about factors:

1. Any integer is a factor of itself.

2. 1 and −1 are factors of *all* integers (except 0).

3. The integer zero has no factors and is not a factor of any integer.

4. A positive integer's largest factor (other than itself) will never be greater than *one half* the value of the integer.

To illustrate these points, consider the integer 30:

There are 16 factors (8 positive, 8 negative) of 30:

$$\{1, -1, 2, -2, 3, -3, 5, -5, 6, -6, 10, -10, 15, -15, 30, -30\}$$

30 is a multiple of (is divisible by) 16 different integers (8 positive, 8 negative).

Prime Numbers

A **prime number** is a positive integer having only *two* positive factors: 1 and the number itself. In other words, a prime number is not divisible by (a multiple of) any positive integer other than itself and 1. *Zero and 1 are not considered prime numbers.*

Recognizing prime numbers is important when simplifying numbers on math exams. For example, if you know that 57 is a prime number, then you won't waste any time trying to reduce $\frac{9}{57}$ to a simpler fraction (a fraction with smaller numbers). Learn to recognize all the prime numbers between 0 and 100, without having to think about it. Here they are:

2	11	23	41	51	61	71	83	91
3	13	29	43	53	67	73	87	93
5	17	31	47	57		79	89	97
7	19	37		59				

Shortcuts for Finding Factors

Determining all the factors of large integers can be tricky; it's easy to overlook some factors. Keep in mind the following rules to help you determine quickly whether one integer is a multiple of (is divisible by) another integer:

If the integer has this feature:	Then it is divisible by:
The integer ends in 0, 2, 4, 6 or 8	2
The sum of the digits is divisible by 3	3
The number formed by the last 2 digits is divisible by 4	4

If the integer has this feature:	Then it is divisible by:
The number ends in 5 or 0	5
The number meets the tests for divisibility by 2 and 3	6
The number formed by the last 3 digits is divisible by 8	8
The sum of the digits is divisible by 9	9

Test these rules by applying them to the numbers between 0 and 100 that are *not* included in the list of prime numbers above.

The Laws of Arithmetic (Combining Numbers and Other Terms)

The **laws of arithmetic** are the fundamental tools we use in combining and simplifying mathematical expressions. Keep in mind the following basic laws of arithmetic, which you will put to use over and over again on your standardized math exam. These rules apply not just to numbers but to variables (such as a and b, or x and y) as well.

Commutative Laws (reversing the order of terms):

rule:	*example:*
$a + b = b + a$	$2 + 3 = 3 + 2$
but $a - b \neq b - a$ (unless $a = b$)	$2 - 3 \neq 3 - 2$
$ab = ba$	$(2)(3) = (3)(2)$
but $\dfrac{a}{b} \neq \dfrac{b}{a}$ (unless $a = b$)	$\dfrac{3}{2} \neq \dfrac{2}{3}$

Associative Laws (grouping the terms together in different ways):

rule:	*example:*
$a + (b + c) = (a + b) + c$	$4 + (5 + 6) = (4 + 5) + 6$
but $a - (b + c) \neq (a - b) + c$	$4 - (5 + 6) \neq (4 - 5) + 6$
$a(bc) = (ab)c$	$(4)(5 \cdot 6) = (4 \cdot 5)(6)$

but $\dfrac{ab}{c} \neq \dfrac{ac}{b}$ or $\dfrac{a}{bc}$ (unless a, b, and c all equal 1) $\dfrac{(4)(5)}{6} \neq \dfrac{(4)(6)}{5}$ or $\dfrac{4}{(5)(6)}$

Distributive Laws (combining one term with each of two or more other terms separately):

rule:	*example:*
$a(b + c) = ab + ac$	$4(5 + 6) = (4)(5) + (4)(6)$
$a - (b + c) = a - b - c$	$4 - (5 + 6) = 4 - 5 - 6$
$\dfrac{a + b}{c} = \dfrac{a}{c} + \dfrac{b}{c}$	$\dfrac{4 + 5}{6} = \dfrac{4}{6} + \dfrac{5}{6}$
but $\dfrac{a}{b + c} \neq \dfrac{a}{b} + \dfrac{a}{c}$	$\dfrac{4}{5 + 6} \neq \dfrac{4}{5} + \dfrac{4}{6}$

At the risk of getting ahead of ourselves, let's take the first distributive law listed above one step further:

rule: $(a + b)(c + d) = ac + ad + bc + bd$

example: $(3 + 4)(5 + 6) = (3)(5) + (3)(6) + (4)(5) + (4)(6)$

In this equation, *a* and *b* are each distributed to *c* and *d*; the result is four separate terms which are added together. If you remember algebra, you may recall the "FOIL" method of combining binomial expressions. (FOIL is an acronym for "*f*irst-*o*uter-*i*nner-*l*ast." Stay tuned for more fun with FOIL in Chapter 10!)

Shortcuts for Combining Strings of Numbers (and other Terms)

When you combine numbers or other mathematical expressions, you must multiply and divide *before* you add and subtract, except where parentheses (or brackets or braces) indicate otherwise. In case you've forgotten this rule, here are four simple examples to refresh your memory:

$$3 \times 4 + 5 = 17 \quad \text{multiply first: } 12 + 5 = 17$$

$$8 - 2 \div 5 \quad \text{divide first: } 8 - \frac{2}{5} = 7\frac{3}{5}$$

$$3 \times (4 + 5) \quad \text{add first: } 3 \times 9 = 27$$

$$(8 - 2) \div 5 \quad \text{subtract first: } 6 \div 5 = \frac{6}{5}$$

These restrictions don't apply, however, when you're dealing *only* with addition and subtraction or *only* with multiplication. In these cases, the associative laws allow you to combine terms in any order you wish. In doing so, group together numbers that are easy to combine. Let's learn how to use each of these four grouping techniques:

- canceling numbers that add up to zero (addition and subtraction)
- combining similarly signed numbers first (addition and subtraction)
- rounding one number up and another one down (addition and subtraction)
- combining numbers when multiplying

Canceling Numbers That Add Up to Zero (Addition and Subtraction)

Quick…combine this string of numbers as indicated:

$$17 - 10 + 11 - 14 + 6 - 7 + 3 = ?$$

The conventional method is to add each term to the next one, from left to right. Here's a quicker way: focus on the one or two numbers having the largest absolute value—in this case, 17 and −14. Look for a combination of smaller numbers in the sequence that cancel out one of these larger numbers. Here's how it works with the string of numbers indicated above:

1. For 17, ask yourself: do two other numbers add up to −17? Yes, −10 and −7.

2. For −14, ask yourself: do two other numbers (other than those you considered already) add up to 14? Yes, 11 and 3.

3. Thus, all terms but the number 6 cancel one another—they add up to zero. So the total sum must be 6.

When you use this shortcut, cross out the numbers with a pencil as you combine them; otherwise, you might lose track of which numbers you've already combined.

Combining Similarly Signed Numbers First (Addition and Subtraction)

Quick…combine this string of numbers as indicated:

$$23 - 12 - 14 + 7 - 8 = ?$$

The conventional method is to subtract 12 from 23, then subtract 14, then add 7, then subtract 8—in other words, to move from left to right. If you use this method when you're in a hurry, it's remarkably easy to overlook a minus sign and to add a number when you should be subtracting it. So why not add together the positive numbers, then add together the negative numbers, then subtract the second total from the first? Here's how it goes:

$$23 + 7 = 30 \qquad \text{combining positive terms}$$

$$12 + 8 + 14 = 34 \qquad \text{combining negative terms}$$

$$30 - 34 = -4 \qquad \text{combining totals}$$

Notice in the second step that the 12 and 8 are added together first, even though the negative numbers are given in a different order (−12...−14...−8). The reason for this is that 12 and 8 add up to a nice round number (20), whereas 12 and 14 do not.

Rounding One Number Up and Another One Down (Addition and Subtraction)

Quick…compute the sum of these four numbers:

$$
\begin{array}{r}
251 \\
423 \\
749 \\
\underline{77} \\
\end{array}
$$

Did you add the digits in the right-hand column, then work your way to the left, as you learned in grade school? There's a quicker way with these four particular numbers. Did you notice that 251 and 749 add up to 1,000—a nice round number. Also, did you notice that 423 and 77 add up to 500—another nice round number. If you did, you could have almost instantaneously determined the total sum: 1,500.

The lesson here is: You can always decrease one number and increase another by the same amount without affecting their sum. So why not do just that if it will make your life easier? 251 exceeds 250 by 1, while 749 is less than 750 by 1. Similarly, 423 is less than 425 by 2, while 77 exceeds 75 by 2. Adding together the four numbers above is the same, then, as adding together these four:

$$250$$
$$425$$
$$750$$
$$\underline{75}$$

Of course, as nice and "round" as these numbers are now, you can still save time by adding them together out of sequence in these three steps:

$$250 + 750 = 1{,}000$$

$$425 + 75 = 500$$

$$1{,}000 + 500 = 1{,}500$$

Let's take this technique a step further, changing our four original numbers:

$$254$$
$$422$$
$$748$$
$$\underline{77}$$

Although 422 and 77 are close to 425 and 75, they don't add up to exactly 500. 422 is *three* less than 425, while 77 is only *two* more than 75. Take the difference of one into account simply by subtracting 1 from 500. Conversely, although you can round off 254 and 748 to 250 and 750, respectively, don't forget to add 2 to the sum, since you rounded down by 4 but up by only 2. Here are the totals: 499 + 1,002 = 1,501.

Multiplying Numbers Out of Sequence

Quick…combine the following numbers:

$$3\tfrac{1}{2} \times 25 \times 16 = ?$$

What order did you choose? You could have performed any of these operations first:

$$3\tfrac{1}{2} \times 25 \qquad 3\tfrac{1}{2} \times 16 \qquad 25 \times 16$$

The last one would probably be your best choice. 25 is one-fourth of 100, so start with 100×16, which equals 1,600, then divide that product by 4 to obtain 400. Nice easy numbers all the way around! Then perform the second operation:

$$400 \times 3\tfrac{1}{2} = (400 \times 3) + (400 \times \tfrac{1}{2}) = 1,400$$

Not too bad. If you start with either of the other two operations, however, the numbers aren't quite as easy to work with:

$$3\tfrac{1}{2} \times 25 = 87\tfrac{1}{2} \qquad\qquad 3\tfrac{1}{2} \times 16 = 48 + 8 = 56$$

$$87\tfrac{1}{2} \times 16 = 1,400 \qquad\qquad 56 \times 25 = 56 \times \frac{100}{4} = \frac{5,600}{4} = 1,400$$

The lesson here is: pair up numbers that are easy to multiply. "Does grouping work for division?" you ask. Well, yes and no. As long as you move from left to right as you group, it works; otherwise, it doesn't. For example, try combining these terms in different sequences:

$$48 \div 12 \div 6$$

You can group 48 with either 12 or 6 first, and the final result will be the same. But if you try dividing 12 by 6 first, look out! You get a different final result.

$$\begin{array}{ccc} (48 \div 12) \div 6 & (48 \div 6) \div 12 & 48 \div (12 \div 6) \\ 4 \div 6 & 8 \div 12 & 48 \div 2 \\ \tfrac{2}{3} & \tfrac{2}{3} & 24 \end{array}$$

Remember: You can add and subtract numbers in any order; you can also multiply (but not divide) a sequence of numbers in any order. So unless you're restricted by parentheses, always look for number pairs that are easy to combine.

Quiz Time

Here are 10 problems to help you determine how well you understand the ideas presented in this chapter. If you have trouble with the easier ones (1–5), go back and review the trouble spots in this chapter. If you can handle the easier ones and the more challenging ones (6–10), consider yourself a "smart test-taker"!

Easier

1. $7 - (-11) + 5 \times 3 - (-23 - 14) =$

2. Without performing division, can you quickly determine the positive factors of 2,496 that are also less than 10?

3. On the real number line, if the distance between x and y is 16.5, which of the following could be the values of x and y?
   (A) −11.5 and 5.5
   (B) 8.25 and −8.75
   (C) 14.5 and 30.5
   (D) −11 and 5.5
   (E) −16.5 and 16.5

4. If x is the sum of all prime factors of 38, and if y is the sum of all prime factors of 84, what is the value of $x - y$?

5. Combine each of these expressions as quickly as possible in your head, using the grouping techniques discussed in this chapter.
   (a) $7 \times 10 + 4 \times 10 + 5$
   (b) $21 - 21 \times 4 - 14 + 12$
   (c) $57 + 31 - 17 + 20 - 3 + 29 - 46$
   (d) $40 - 21 + 37 - 16 - 19 - 2$
   (e) $16 \times 13 \times \frac{1}{2} - 7 \times \frac{3}{2} \times 18$

More Challenging

6. If n is a positive even integer, and if $n \div 3$ results in a quotient with a remainder of 1, how many of the following expressions are divisible by 3?

$n - 1$ $n + 1$ $n + 2$ $n \cdot 2$ $n \cdot 3$

7. If x and y are negative integers, and if $x - y = 1$, what is the least possible value of xy?

8. Assume that the symbol (*) represents a digit in the five-digit number 62,*79 that is a multiple of 3. If the sum of the number's five digits is divisible by 4, what is the value of *?

9. If x, y, and z are consecutive negative integers, and if $x > y > z$, which of the following must be a positive odd integer?

(A) xyz
(B) $x + y + z$
(C) $x - yz$
(D) $x(y + z)$
(E) $(x - y)(y - z)$

10. Combine each of these expressions as quickly as possible in your head, using the grouping and rounding techniques discussed in this chapter.

(a)	(b)	(c)	(d)	(e)
81	327	166	407	609
19	+623	89	1143	29
33		214	557	+361
+17		+151	+8993	

Answers and Explanations

1. The answer is 70. First multiply $5 \cdot 3$. Then combine the terms in parentheses. Then combine the resulting terms (remember: subtracting a negative number is equivalent to adding a positive one). Here are the steps:

$7 - (-11) + 5 \times 3 - (-23 - 14) =$

$7 - (-11) + 15 - (-37) =$

$7 + 11 + 15 + 37 = 70$

2. The positive factors of 2,496 that are also less than 10 include: 2, 3, 4, 6, and 8.

3. The correct answer choice is (D).

The difference between -11 and 5.5 is 16.5.

4. The answer is 9.

The prime factors of 38 include: $\{2, 19\}$. The sum of these numbers is 21.

The prime factors of 84 include: $\{2, 3, 7\}$. The sum of these numbers is 12.

$21 - 12 = 9$.

Thus, $x - y = 9$.

5. Here are the answers to the five problems:

(a)   115

(b)   −65

(c)   71

(d)   0

(e)   −85

6. The answer is three. Start with 2, then 4, then 6, and so forth (positive even integers), as the value of n. Test each value in turn. You'll find that only the numbers in the following sequence leave a remainder of 1 when divided by 3: $\{4, 10, 16, \ldots\}$. (Notice that the numbers increase by 6 in sequence.) Next, try a few of these numbers as the value of n in each of the five expressions. You'll find that three of the expressions—$n - 1$, $n + 2$ and $n \cdot 3$—are divisible by 3. The other two expressions are not.

7.   The answer is 2. Use values for x and y with small absolute values (numbers approaching zero) in order to obtain the smallest product when you multiply them together. The two smallest values that satisfy the equation are: $y = -2$, and $x = -1$. Accordingly, the least possible value of xy is 2.

8.   The answer is 0. If the sum of the digits of a number is divisible by 3, the number is also divisible by 3. The sum of the digits in the number 62,*79, excluding *, is 24. Thus, if the number is a multiple of (is divisible by) 3, * must equal either 0, 3, 6 or 9. Taking into account separately that the sum of the digits is divisible by 4, * must equal either 0, 4 or 8. The only common number under the two tests is 0 (zero). Thus, * = 0.

9.   The correct answer choice is (E). Given that x, y, and z are consecutive negative integers, either one integer is odd or two integers are odd. Also, a negative number multiplied by a negative number yields a positive number. With this in mind, consider each answer choice in turn:

(A) xyz must be negative and even.

(B) $x + y + z$ must be negative; whether (B) is odd or even depends on whether one or two of the three integers are odd.

(C) $x - yz$ must be negative (since yz, a positive number, is subtracted from the negative number x). Whether (C) is odd or even depends on whether one or two of the three integers are odd. yz must be even; however, x could either be even or odd.

(D) $(y + z)$ must be negative and odd. Thus, the product of x and $(y + z)$ must be positive (either odd or even).

(E) $(x - y)$ must be odd and positive, since $x > y$. Similarly, $(y - z)$ must be odd and positive, since $y > z$. The product of the terms must therefore be odd and positive.

10. The answers to the five problems are:

(a)   150

(b)   950

(c)   620

(d)   11,100

(e)   999

4

Fractions, Ratios,
and Proportion

I t's no coincidence that we're going to look at fractions, ratios, and proportion together in the same chapter. That's because, when it comes right down to it, these three concepts are essentially the same. If you don't agree, I promise that I'll make a convert out of you by the time you've finished this chapter. Most of the examples used througout this chapter involve numbers rather than variables (like *x* and *y*). We're going easy on the variables because they smack of the dreaded "A" word (algebra), and we're putting off algebra as much as possible until Chapter 10. **Keep in mind**, however, that the discussion throughout this chapter applies to variables as well as numbers.

Fractions

The world wouldn't be the same without fractions. At football games, *half-time* would be known instead as "a brief respite." Instead of saying, "I'll be home in *half an hour*," you'd have to say, "I'll be home when the minute hand has moved 180 degrees." You wouldn't know how fast you were driving in *miles per hour* when the authorities stop you for speeding. And people who design math tests would be in a terrible fix! Fractions are everywhere, on math tests as in real life.

In the first part of this chapter, **you'll learn** all the tricks for simplifying and reducing fractions as well as combining fractions—with addition, subtraction, multiplication, and division. You'll need the skills you learn here to handle more complex math problems—especially those involving algebra.

The Many Faces of Fractions

Let's start by looking at different kinds of fractions. As we do so, don't worry too much about the terminology; pay closer attention to the concepts. A *fraction* is a number (or other term) of the form $\frac{x}{y}$ (where $y \neq 0$). x is referred to as the *numerator*, and y is referred to as the *denominator*. x and y themselves can be numbers—integers or non-integers—or any other mathematical expression. The first fraction below is a *simple fraction*, while each of the other two is a *complex fraction*:

$$\frac{13}{6} \qquad\qquad \frac{\frac{2}{3}}{x} \qquad\qquad \frac{\frac{13}{6}}{a - \frac{5}{2}}$$

Let's focus first on simple fractions with numbers (like the fraction on the left). This type of fraction can be either a *proper* fraction, *improper* fraction, or *mixed number*. Here's the difference among the three types, along with examples of each:

$\frac{6}{47}$	$-\frac{8}{17}$	proper fractions	value of fraction is less than 1 (denominator is greater than numerator)
$\frac{47}{6}$	$-\frac{17}{8}$	improper fractions	value of fraction is greater than 1 (numerator is greater than denominator)
$7\frac{5}{6}$	$-2\frac{1}{8}$	mixed numbers	improper fraction expressed as integer with proper fraction remainder

Converting Mixed Numbers to Fractions (and Vice Versa)

Did you recognize in the preceding examples that $\frac{47}{6} = 7\frac{5}{6}$ and that $-\frac{17}{8} = -2\frac{1}{8}$? Any improper fraction can be expressed instead as a mixed number (and vice versa). To convert an improper fraction to a mixed number:

1. Divide the numerator by the denominator.

2. The number of complete times the denominator "goes into" the numerator becomes the integer of the mixed number.

3. The number that's left over—the remainder—becomes the numerator of the fraction that's left over (the denominator stays the same).

$$\frac{47}{6} = 7\frac{5}{6}$$

6 goes into 47 seven complete times.

$6 \times 7 = 42$, leaving a remainder of 5 of the original 47 sixths.

$$-\frac{17}{8} = -2\frac{1}{8}$$

8 goes into 17 two complete times.

$8 \times 2 = 16$, leaving a remainder of 1 of the original 17 eighths.

To convert a mixed number to an improper fraction, multiply the denominator by the integer, then add the product to the numerator, and place the sum over the original denominator:

$$7\frac{5}{6} = \frac{(6)(7) + 5}{6} = \frac{47}{6}$$

$$-2\frac{1}{8} = -\frac{(8)(2) + 1}{8} = -\frac{17}{8}$$

Negative Signs and Fractions

In the preceding section, three of the six examples we used were negative numbers:

$$-\frac{8}{17} \qquad\qquad -\frac{17}{8} \qquad\qquad -2\frac{1}{8}$$

Notice that the negative or "minus" sign (–) in each number *precedes* the fraction; the sign is neither part of the numerator nor part of the denominator. What would happen to the first two numbers above if we placed the negative sign in the fraction?

$$-\frac{8}{7}, \; \frac{-8}{7}, \; \frac{8}{-7} \qquad\qquad -\frac{17}{8}, \; \frac{-17}{8}, \; \frac{17}{-8}$$

Does the value of the fraction change by moving the negative sign? No. We've simply indicated three different ways of expressing the same value—the same spot on the real number line. The first method is probably the least confusing; that's why mathematicians (and people who design math tests) almost always put negative signs *outside* fractions, except where other terms are included within the fraction; for example:

$$\frac{x-3}{4}$$

Comparing Fractions by the "Cross-Product" Method

Quick...which of these two fractions is larger in value?

$$\frac{7}{9} \qquad\qquad \frac{3}{4}$$

Hmm, not so easy, is it? Here's a fool-proof way to compare the relative sizes of two fractions. Multiply the numerator of each fraction by the other fraction's denominator. Compare these two products. The fraction whose numerator contributes to the larger product is the greater of the two fractions. Comparing the cross products for the fractions above, 28 (7 · 4) is greater than 27 (9 · 3). Accordingly, $\frac{7}{9}$ is greater than $\frac{3}{4}$.

Reducing a Fraction to its "Lowest Terms"

A fraction reduced to its "lowest terms" is one that uses the smallest possible integers without changing the value of the overall fraction itself. Here are two examples:

$$\frac{21}{14} = \frac{3}{2} \longleftarrow \text{ lowest terms}$$

$$-\frac{22}{44} = -\frac{1}{2} \longleftarrow \text{ lowest terms}$$

Why bother to change fractions to their lowest terms? First, on standardized exams, correct answer choices are almost always expressed in lowest terms. Second, it's easier to work with small numbers than big ones. To reduce a fraction to its lowest terms, look for factors common to its numerator and denominator. Always look for the largest possible factors. Then, either mentally or with paper and pencil, separate out those factors—or "factor them out"—by expressing the original number as a product or multiple of other numbers. Let's look at three fractions to see how it works:

$$\frac{64}{96} = \frac{(8)(8)}{(8)(12)} \qquad \text{8 factored out}$$

$$\frac{28}{42} = \frac{(7)(4)}{(7)(6)} \qquad \text{7 factored out}$$

$$\frac{18}{27} = \frac{(9)(2)}{(9)(3)} \qquad \text{9 factored out}$$

Now that we've "factored out" a number common to both the numerator and denominator, we can "cancel" the number—in other words, cross it out or ignore it. Why? Simply because any number divided by itself equals 1. Here's the result:

$$\frac{64}{96} = \frac{(8)(8)}{(8)(12)} = \frac{8}{12} \qquad \frac{28}{42} = \frac{(7)(4)}{(7)(6)} = \frac{4}{6} \qquad \frac{18}{27} = \frac{(9)(2)}{(9)(3)} = \frac{2}{3}$$

We're not quite done, however. In the first two examples, the resulting fractions are still not expressed their lowest terms. You can "factor out" and "cancel" yet another integer:

$$\frac{8}{12} = \frac{(4)(2)}{(4)(3)} = \frac{2}{3} \qquad\qquad \frac{4}{6} = \frac{(2)(2)}{(2)(3)} = \frac{2}{3}$$

Notice that in the first two examples we could have factored out a 32 and a 14 respectively, in the first step, avoiding a second step:

$$\frac{64}{96} = \frac{(32)(2)}{(32)(3)} = \frac{2}{3} \qquad\qquad \frac{28}{42} = \frac{(14)(2)}{(14)(3)} = \frac{2}{3}$$

If you happen to recognize the largest common factor, that's great! With larger numbers, however, it's easy to overlook the largest common factor (for example, 14 in the second fraction above). That's okay, though; you'll get to the simplest form anyway—probably in a quick second step.

Remember: In reducing a fraction to its lowest terms, always look for the largest factor common to the numerator and denominator. But don't spend too much time looking for it; smaller common factors will also get you there eventually.

Adding and Subtracting Fractions Which Have the Same Denominator

Remember the distributive law for division from Chapter 3? Well, it's back! Just for the record, here's the law again:

$$\frac{a+b}{c} = \frac{a}{c} + \frac{b}{c}$$

To add or subtract two fractions having the *same denominator*, you just apply the distributive law from right to left. Add together the two numerators, and place the sum over the same (original) denominator:

$$\frac{3}{7}+\frac{5}{7}=\frac{8}{7} \qquad\qquad \frac{8}{5}-\frac{11}{5}=-\frac{3}{5}$$

Finding the Lowest Common Denominator (LCD)

As long as denominators are the same, adding or subtracting fractions is no big deal. However, adding or subtracting fractions having *different denominators* isn't quite as simple. Here's what you need to do:

1. Find a "common denominator," which is an integer that is a multiple of each denominator.

2. For each fraction, multiply the numerator and denominator by an integer that will result in that common denominator.

How do you find a common denominator? A real easy way is to multiply the denominators together. The product will be a common multiple. Here are two examples:

$$-\frac{13}{4}+\frac{8}{3}=-\frac{13(3)}{4(3)}+\frac{8(4)}{3(4)}=-\frac{39}{12}+\frac{32}{12}=\frac{-39+32}{12}=-\frac{7}{12}$$

(12 is a multiple common to the denominators 4 and 3)

$$\frac{5}{6}-\frac{1}{4}=\frac{5(4)}{6(4)}-\frac{1(6)}{4(6)}=\frac{20}{24}-\frac{6}{24}=\frac{20-6}{24}=\frac{14}{24}\text{ or }\frac{7}{12}$$

(24 is a multiple common to the denominators 6 and 4)

$$\frac{20}{24}-\frac{6}{24}\qquad \frac{20-6}{24}\quad \frac{14}{24}$$

In the first example, 12 also happens to be the smallest common multiple, or *least common denominator (LCD)*, of 4 and 3. In other words, 12 is the smallest positive integer that is a multiple of both 4 and 3. The second example, however, presents a different situation: 12, not 24, is the LCD of 6 and 4. We got the right answer by using a larger multiple (24), but we had to deal with larger numbers along the way. Look what else we had to deal with: $\frac{14}{24}$ was not a fraction expressed in lowest terms. We had to factor out (and cancel) 2 from both numerator and denominator to reduce the fraction to its simplest form: $\frac{7}{12}$. So we got to the LCD; it just took one more step because we didn't choose the LCD from the outset. Let's try the second example again, using the LCD (12) instead of 24 as our common denominator:

$$\frac{5}{6} - \frac{1}{4} = \frac{5(2)}{6(2)} - \frac{1(3)}{4(3)} = \frac{10}{12} - \frac{3}{12} = \frac{10-3}{12} = \frac{7}{12}$$

We've saved a step and we worked with smaller (easier) numbers along the way.

Everything we've talked about so far regarding common denominators also applies to adding or subtracting *more than two* fractions. In the following example, which involves three fractions, the LCD is 36:

$$\frac{5}{6} + \frac{7}{9} - \frac{3}{4} = \frac{30}{36} + \frac{28}{36} - \frac{27}{36} = \frac{30+28-27}{36} = \frac{31}{36}$$

Shortcuts for Finding the LCD

How do you know for sure if you've chosen the *least* common denominator to combine your fractions? There are two useful tips you should keep in mind. First, if one of two denominators is 2, it's easy: If the other number is *even*, then the lowest common denominator is the other number.

even pair	lowest common denominator
2, 4	4
2, 12	12
2, 86	86

If the other number is *odd*, then the lowest common denominator is *twice* the other number.

odd pair	lowest common denominator
2, 5	10
2, 11	22
2, 49	98

Here's the second tip. Always check to see if the largest denominator is a multiple of *each* of the other(s). If so, the largest denominator is the LCD.

denominators	lowest common denominator
7, 21	21
3, 4, 24	24
9, 12, 36	36
2, 3, 5, 30	30

If the problem doesn't fall into one of these patterns, however, some systematic trial-and-error may be necessary to find the LCD. Try multiplying together different combinations of denominators, then check to see whether the product is also a multiple of the remaining denominators. Keep in mind that in many cases the LCD will simply be all denominators multiplied together:

denominators	LCD	how we found the LCD
2, 6, 11	66	product of two of the numbers (6 · 11)
6, 8, 9	72	product of two of the numbers (8 · 9)
3, 4, 7	84	product of all three numbers (3 · 4 · 7)

Multiplying Fractions Together

To multiply one fraction by another, multiply the two numerators and multiply the two denominators (in this case, the denominators need not be the same):

$$\left(\frac{13}{2}\right)\left(-\frac{3}{7}\right) \ = \ -\frac{(13)(3)}{(2)(7)} \ = \ -\frac{39}{14}$$

To simplify the multiplication, always look for the possibility of canceling factors common to *either* numerator and *either* denominator:

$$\frac{9}{7}\cdot\frac{7}{6} \ = \ \frac{{}^3\cancel{9}\cdot\cancel{7}^1}{{}_1\cancel{7}\ \cancel{6}_2} \ = \ \frac{3}{2}$$

Using the "Reciprocal" to Divide Numbers That Include Fractions

Division problems that involve fractions can be expressed in two different ways; for example:

$$\frac{2}{3} \div \frac{11}{5} \quad \text{is the same as} \quad \frac{\frac{2}{3}}{\frac{11}{5}}$$

In either case, to divide one fraction by another, multiply the first fraction (the numerator fraction) by the *reciprocal* of the second one (the denominator fraction). The reciprocal is the same fraction, but with the numerator and denominator exchanged or "flipped." (The product of a number and its reciprocal is 1.) Here's how it works with our example:

$$\frac{\frac{2}{3}}{\frac{11}{5}} = \frac{2}{3} \cdot \frac{5}{11} = \frac{10}{33} \qquad \text{The reciprocal of } \frac{11}{5} \text{ is } \frac{5}{11}$$

If the denominator is not already a fraction itself, express it as such by placing it in a numerator over the denominator 1, then flip it:

$$\frac{\frac{5}{9}}{15} = \frac{\frac{5}{9}}{\frac{15}{1}} = \frac{5}{9} \cdot \frac{1}{15} = \frac{\cancel{5}^{1}}{9} \cdot \frac{1}{\cancel{15}_{3}} = \frac{1}{27} \qquad 15 \text{ is the same as } \frac{15}{1}$$

$$\text{The reciprocal of } \frac{15}{1} \text{ is } \frac{1}{15}$$

Always try to cancel common factors *before* multiplying together the two fractions.

Remember: it's easier to work with small numbers than with large ones, and you'll probably have to have to cancel common factors eventually in any event.

How can we get away with simply flipping the denominator fraction and multiplying the result by the numerator? Well, here's what's really going on behind the scenes (in the first example):

$$\frac{\frac{2}{3}}{\frac{11}{5}} = \frac{\frac{2}{3}\left(\frac{5}{11}\right)}{\frac{11}{5}\left(\frac{5}{11}\right)} = \frac{\frac{2}{3}\left(\frac{5}{11}\right)}{1} = \frac{2}{3}\cdot\frac{5}{11}$$

As you can see, we're actually multiplying both the numerator and the denominator of the complex fraction by the same number: 5/11. In doing so, we are aren't changing the value of the complex fraction at all. So, "flipping" the denominator is really just a shortcut to a multi-step process.

Remember: Dividing one number by another is the same as multiplying the first number by the reciprocal of the second number.

Ratios

A *ratio* is an expression of proportion or comparative size. Let's put this in a verbal context. Given 12 males and 16 females, all of the following are true:

- the male/female ratio is 12:16, 3:4, $\frac{12}{16}$, or $\frac{3}{4}$

- the ratio of males to females is 12 to 16, or 3 to 4

- the number of males is $\frac{12}{16}$ (or $\frac{3}{4}$) the number of females

- the number of males multiplied by $\frac{16}{12}$ (or $\frac{4}{3}$) equals the number of females

- the female/male ratio is 16:12, 4:3, $\frac{16}{12}$, or $\frac{4}{3}$

- the ratio of females to males is 16 to 12, or 4 to 3

- the number of females is $\frac{16}{12}$ (or $\frac{4}{3}$) the number of males

- the number of females multiplied by $\frac{12}{16}$ (or $\frac{3}{4}$) equals the number of males

The order of the terms of a ratio is very important. For instance, in the foregoing example, the ratio of males to females is 12 to 16, *not* 16 to 12.

As suggested here, the ratio of two numbers (x to y) can be expressed as either "$x{:}y$," "x to y," or "x/y". Ratios, just like fractions, can be reduced to lowest terms by canceling common factors. Notice that in these examples, we changed 12:16 to 3:4 and 16:12 to 4:3.

Ratios Are Fractions in Disguise

Did you notice in the preceding section that all of the ratios were expressed as fractions as well. That's because when it comes down to it, ratios are no different than fractions. In fact, every ratio is a fraction, and every fraction is a ratio. They are just two different ways of comparing the size of two amounts—of expressing relative size or *proportion.*

Let's take this idea a step further. In any ratio—or in any fraction—the size of either term considered alone is not important; the key is their size relative to each other. In other words, we can use numbers of any size, as long as the relative size (or proportion) of the numbers remains the same:

$$12{:}16 = 3{:}4 = .084 : .112$$ 　　same ratio with different numbers

$$\frac{12}{16} = \frac{3}{4} = \frac{.084}{.112}$$ 　　same value with different numerator/denominator pairs

The "Pie" Approach to Understanding Ratios

Perhaps the best way to think about a ratio is as a whole made up of different parts—like a whole pie divided into pieces. Let's use our male-to-female ratio of 12:16 (3:4) as an example. The "whole pie" is the total number of people. This whole is cut into two slices—males and females. We can change the total number of people without affecting the proportions of the slices—that is, without affecting the male-female ratio:

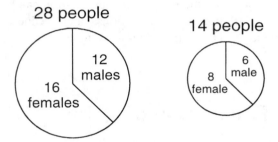

In either pie above, notice that the "whole" (total number) equals the sum of its "slices" (number of males plus number of females). By thinking about the ratio in this way, you can easily determine the size of each slice as a portion of the whole:

- The portion of all students who are males is 12:28 or 3:7. In other words, $\frac{3}{7}$ of all students are males.

- The portion of all students who are females is 16:28 or 4:7. In other words, $\frac{4}{7}$ of all students are females.

Essentially, you're adding up the fractional slices to equal the number 1 (one whole pie):

$$\frac{3}{7} + \frac{4}{7} = \frac{7}{7} \text{ (the whole pie)}$$

This approach works just as well for pies that are cut into *more than two* slices. Assume, for example, that three lottery winners—X, Y and Z—are sharing a lottery jackpot of $195,000. However, their shares are not equal; X's share is 1/5 of Y's share and 1/7 of Z's share. How would you determine the dollar amount of each winner's shares?

By thinking about these shares as slices of a whole pie—the whole jackpot. Here's how it works:

First, express the winners' proportionate shares as ratios. The ratio of X's share to Y's share is 1 to 5. Similarly, the ratio of X's share to Z's share is 1 to 7. The jackpot share ratio as follows:

$$X : Y : Z = 1 : 5 : 7$$

X's winnings account for 1 of 13 equal parts (1 + 5 + 7) of the total jackpot. 1/13 of $195,000 is $15,000. Accordingly, Y's share is 5 times that amount, or $75,000, and Z's share is 7 times that amount, or $105,000.

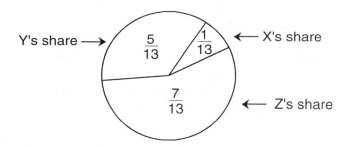

Proportion—a Taste of Algebra

Fractions and ratios inherently involve the concept of proportion. It's the relative size (or proportion) of the numbers, not the numbers considered individually, that is important. Think about it in this way:

- What happens when you change the value of a numerator or denominator? Unless you change the other part (numerator or denominator) of the fraction *proportionately*, you've altered the value of the fraction as a whole!

- What happens when you change one of the numbers in a ratio? Unless you change the other number(s) *proportionately*, you've altered the ratio!

This begs the real question: If you change one of the numbers, how do you go about changing the other number(s) proportionately? Well, what you *can't* do is simply add or subtract the same number from each of the terms. This would be an *equal* change, but not necessarily a *proportionate* change:

$$\frac{3}{4} \neq \frac{3+1}{4+1} \qquad\qquad 3:4 \neq 4:5$$

Instead, set up a simple equation, using a variable ("*x*" is always a tasteful choice, but we'll be a bit gauche and use a "?" here) to represent the unknown value *after* the change:

$$\frac{3}{4} = \frac{3+1}{?}$$

For all you "verbal" folks, what's happening here is that we're asking the question

"3 is to 4 as 4 is to what?"

We're just expressing this question as an equation. It's beginning to look a lot like algebra, isn't it? At the risk of getting ahead of ourselves (algebra is the main course in Chapters 10–12), solve for the unknown value by isolating "?" on one side of the equation:

$$(4)(?)\frac{3}{4} = \frac{4}{?}(4)(?) \qquad \text{multiply both sides of the equation by (4)(?)}$$

$$(3)(?) = 16 \qquad \text{cancel terms}$$

$$? = \frac{16}{3} \text{ or } 5\frac{1}{3} \qquad \text{divide both sides of the equation by 3}$$

Thus, the ratio 3:4 is equivalent to (the same thing as) $4 : \frac{16}{3}$.

Here's some good news: If the numbers involved are simple enough, you don't have to resort to algebra. Ask yourself what number you multiplied the original one by to get the new number, then multiply the other terms by the same number. In the foregoing example, we increased 3 to 4—the same as multiplying 3 by $1\frac{1}{3}$. So we can maintain a constant ratio by also multiplying 4 by $1\frac{1}{3}$. (The result is $\frac{16}{3}$, which is no coincidence!)

If we had increased 3 to 6 (instead of 4), the math would have been even easier. We'd be multiplying 3 by 2 to obtain 6, so we would multiple 4 by 2 as well:

$$\frac{3}{4} = \frac{3 \cdot 2}{4 \cdot 2} = \frac{6}{8}$$

On math tests, questions involving ratio and proportion often appear as "word" problems. We'll be dealing more with these kinds of problems in Chapter 12.

Remember: You can multiply (or divide) each term of a ratio (or fraction) by the same number without changing the overall ratio. But you can't add (or subtract) the same number to (from) one term in a ratio and expect the ratio to remain unchanged.

Quiz Time

Here are 10 problems to test your skill in applying the concepts in this chapter. After attempting all 10 problems, read the explanations that follow. Then go back to the chapter and review your trouble spots. If you can handle the easier problems *and* the more challenging ones, consider yourself a "smart test-taker"!

Easier

1. Convert expressions (a) and (b) to simple fractions. Convert expressions (c) and (d) to mixed numbers.

    (a) $13\frac{2}{7}$

    (b) $-5\frac{1}{3}$

    (c) $\frac{47}{4}$

    (d) $-\frac{19}{9}$

2. Multiply together the fractions in each of groups (a) through (d) below. Express your answer in lowest terms.

    (a) $\frac{1}{4} \cdot \frac{2}{5} \cdot \frac{5}{6}$ (b) $\frac{11}{3} \cdot \frac{2}{7}$ (c) $\left(-\frac{36}{9}\right) \cdot \left(-\frac{15}{28}\right)$ (d) $\frac{99}{44} \cdot \frac{17}{2} \cdot \frac{2}{9} \cdot \left(-\frac{5}{34}\right)$

3. If $\dfrac{a}{b} \cdot \dfrac{b}{c} \cdot \dfrac{c}{d} \cdot \dfrac{d}{e} \cdot x = 1$, then $x =$

    (A) $\dfrac{a}{e}$ (B) $\dfrac{e}{a}$ (C) e (D) $\dfrac{1}{a}$ (E) $\dfrac{be}{a}$

4. Reduce $\dfrac{4\frac{1}{2}}{4\frac{1}{8} - 3\frac{2}{3}}$ to a simple fraction in lowest terms.

5. If $a \neq 0$ or 1, then $\dfrac{\dfrac{1}{a}}{2 - \dfrac{2}{a}}$ is equivalent to which of the following fractions?

(A) $\dfrac{1}{2a - 2}$ (B) $\dfrac{2}{a - 2}$ (C) $\dfrac{1}{a - 2}$ (D) $\dfrac{1}{a}$ (E) $\dfrac{2}{2a - 1}$

More Challenging

6. The denominator of a certain fraction is twice as large as the numerator. If 4 is added to both the numerator and denominator, the value of the new fraction is $\dfrac{5}{8}$. What is the denominator of the original fraction?

7. Machine X, Machine Y and Machine Z each produce widgets. Machine Y's rate of production is 1/3 that of Machine X, and Machine Z's production rate is twice that of Machine Y. If Machine Y can produce 35 widgets per day, how many widgets can the three machines produce per day working simultaneously?

8. There is enough food at a picnic to feed either 20 adults or 32 children. All adults eat the same amount, and all children eat the same amount. If there are 15 adults at the picnic, and if all 15 adults are fed, how many children can still be fed?

9. A family of two adults and two children is going together to the local zoo, which charges exactly twice as much for each adult's admission ticket as for each child's admission ticket. If the total admission price for the family of two adults and two children is $12.60, what is the price of a child's ticket?

10. An animal shelter houses only two different types of animals—dogs and cats. If d represents the number of dogs, and if c represents the number of cats, which of the following expresses the portion of animals at the shelter that are dogs?

(A) $\dfrac{d + c}{d}$ (B) $\dfrac{c + d}{c}$ (C) $\dfrac{d}{c}$ (D) $\dfrac{c}{d}$ (E) $\dfrac{d}{d + c}$

Answers and Explanations

1. Here are the answers to the four problems:

   (a) $\frac{93}{7}$

   (b) $-\frac{16}{3}$

   (c) $11\frac{3}{4}$

   (d) $-2\frac{1}{9}$

2. Here are the answers to the four problems:

   (a) $\frac{1}{12}$

   (b) $\frac{22}{21}$

   (c) $\frac{15}{7}$

   (d) $-\frac{5}{8}$

3. The correct answer choice is (B).

   All variables except a and e cancel out:

   $$\frac{a}{\cancel{b}} \cdot \frac{\cancel{b}}{\cancel{c}} \cdot \frac{\cancel{c}}{\cancel{d}} \cdot \frac{\cancel{d}}{e} \cdot x = 1$$

   To isolate x on one side of the equation, multiply both sides by $\frac{e}{a}$:

   $$\left(\frac{e}{a}\right)\left(\frac{a}{e}\right)(x) = (1)\left(\frac{e}{a}\right)$$

   $$x = \frac{e}{a}$$

4. The answer is $\frac{108}{11}$. First, convert all mixed numbers into simple fractions:

   $$\frac{\dfrac{9}{2}}{\dfrac{33}{8} - \dfrac{11}{3}}$$

Next, combine the two fractions in the denominator over their LCD, which is 24:

$$\dfrac{\dfrac{9}{2}}{\dfrac{99}{24} - \dfrac{88}{24}} = \dfrac{\dfrac{9}{2}}{\dfrac{99 - 88}{24}} = \dfrac{\dfrac{9}{2}}{\dfrac{11}{24}}$$

To eliminate the fractional denominator, multiply the numerator by the denominator's reciprocal. Then simplify by factoring:

$$\dfrac{9}{2} \cdot \dfrac{24}{11} = \dfrac{9}{1} \cdot \dfrac{12}{11} = \dfrac{108}{11}$$

5. The correct answer choice is (A).

Simplify this complex fraction by multiplying *every* term by a:

$$\dfrac{\dfrac{1(a)}{a}}{2(a) - \dfrac{2(a)}{a}} = \dfrac{1}{2a - 2}$$

6. The answer is 12. Represent the original fraction by $\dfrac{x}{2x}$, and add 4 to both the numerator and denominator:

$$\dfrac{x + 4}{2x + 4} = \dfrac{5}{8}$$

$8x + 32 = 10x + 20$ (cross multiplication)

$$12 = 2x$$

$$x = 6$$

The original denominator is $2x$, or 12.

7. The answer is 210.

The ratio of X's rate to Y's rate is 3 to 1, and the ratio of Y's rate to Z's rate is 1 to 2. The ratio among all three can be expressed as 3:1:2 (X:Y:Z). Accordingly, Y's production accounts for 1/6 of the total widgets that can be produced per day by all three machines. Given that Y can produce 35 widgets per day, all three machines can produce (35)(6), or 210, widgets per day.

8. The answer is 8.

Since 20 adults would eat all the food, 15 of 20 adults will eat 15/20, or 3/4, of the food. That leaves 1/4 of the food. Since all of the food would feed 32 children, 1/4 of the food will feed 1/4 of 32, or 8, children.

9. The answer is $2.10.

Think of $12.60 as the total "pie." The ratio of ticket prices among the four family members is 1:1:2:2. These numbers add up to 6; thus, each child's ticket costs 1/6 of the $12.60.

10. The correct answer choice is (E).

The shelter houses $(d + c)$ animals altogether. Of these animals, d are dogs. That portion can be expressed as the fraction $\dfrac{d}{d + c}$.

5

Decimals
and Percents

Okay, you've read four of the fifteen chapters in this book; so you're 4/15 of the way through it. Quick...how far are you in terms of a *decimal number* or a *percentage*? If the number of chapters were increased by 30%, then how far would you be at this point in terms of a fraction, decimal number, or percentage? These are just the kinds of questions **you'll learn** to respond to quickly and accurately in this chapter.

A truly **smart test-taker** is "trilingual" in a mathematical sense, equally adept in handling fractions, decimal numbers, and percents, and easily able to convert (or "translate") one form to another. We learned how to deal with fractions in Chapter 4. **Now let's** round out your *form*–al education with decimals and percents.

Decimals

Any real number can be expressed in decimal form. The value of each digit in a decimal number depends on its position (place) relative to the decimal point. For example, in the number 938.421:

- 9 is the "hundreds" digit, and it has a place value of 100 (3 places to the left of the decimal point)

- 3 is the "tens" digit, and it has a place value of 10 (2 places to the left of the decimal point)

- 8 is the "ones" digit, and it has a place value of 1 (1 places to the left of the decimal point)

- 4 is the "tenths" digit, and it has a place value of $\frac{1}{10}$ (1 places to the left of the decimal point)

- 2 is the "hundredths" digit, and it has a place value of $\frac{1}{100}$ (2 places to the left of the decimal point)

- 1 is the "thousandths" digit, and it has a place value of $\frac{1}{1,000}$ (3 places to the left of the decimal point)

Accordingly:

$$938.421 =$$

$$9(100) + 3(10) + 8(1) + 4\left(\frac{1}{10}\right) + 2\left(\frac{1}{100}\right) + 1\left(\frac{1}{1000}\right) =$$

$$900 + 30 + 8 + \frac{4}{10} + \frac{2}{100} + \frac{1}{1000}$$

Notice that the words used to describe digits to the left of the decimal point do not match those to the right. For example, while the "hundreds" digits is the *third* one to the left of the decimal point, the "hundredths" digit is the *second* one to the right of the decimal point. That's because there is no "units" digit to the right of the decimal point.

Dealing with Decimals (Addition and Subtraction)

Adding or subtracting numbers that include decimal points (we'll refer to such numbers here as "decimal numbers") is really pretty straightforward. The best way to do it is to arrange the numbers in columns with the decimal points (as well as all corresponding digits) lined up vertically. Here's a rather extreme example to help illustrate the point:

$$
\begin{array}{r}
234.01 \\
4.3 \\
1967. \\
+\ \ \ \ .867 \\
\hline
2206.177
\end{array}
$$

Notice that all of the "units" digits are lined up vertically; and the same applies to all other corresponding digits as well. Next, add (or subtract) the digits in each column, carrying numbers to the next column as needed. We won't go into the details, since this is pretty basic stuff you probably already know. There is one bit of advice, though, that's well worth emphasizing here:

Don't try to add or subtract decimal numbers in your head. It's too easy to confuse the digits this way, especially when you're in a hurry. Use your pencil, and line up those numbers in columns on your paper!

Dealing with Decimals (Multiplication and Division)

Do you freeze up when multiplying or dividing decimal numbers, even though you don't have any trouble adding or subtracting them? If so, thaw out by applying a few simple shortcuts for dealing with decimals. Consider these two problems:

$$
\begin{array}{r}
23.6 \\
\times\ .07 \\
\end{array}
\qquad\qquad
11.49 \div .003 \ \ \text{or} \ \ \dfrac{11.49}{.003}
$$

Wouldn't it be nice if those little dots (the decimal points) would just go away. Well, go ahead and ignore them. You'll still end up with the correct answer, as long as you follow a few simple steps.

Multiplication

Here are the steps for multiplication problems:

1. Remove (or just ignore) all "leading" zeros—the ones just to the right of the decimal point

2. Remove (or just ignore) all the decimal points.

3. Multiply together the "stripped-down" numbers—whatever is left after you strip away leading zeros and decimal points.

4. Count the *total* number of digits to the right of all decimal points in your original problem.

5. Insert a decimal point in your resulting product so that the number of decimal places equals that total.

Here's how it works with the multiplication example above:

$23.6 \times .07$	original expression (3 decimal places altogether)
236×7	decimals and "leading" zero removed
1652	stripped-down numbers multiplied together
1.652	decimal point inserted (3 decimal places)

Remember: Whenever you multiply numbers together, the number of decimal places (digits to the right of the decimal point) in the resulting product will be the same as the total number of decimal places in the original numbers.

Division

The steps for dealing with decimals in division are similar to those for multiplication, but the final step is a bit trickier. Just as you did with multiplication:

1. Remove (or just ignore) all "leading" zeros—the ones just to the right of the decimal point

2. Remove (or just ignore) all the decimal points.

3. Perform division on the "stripped-down" numbers—whatever is left after you strip away leading zeros and decimal points.

Here's where it gets tricky. To determine where to insert the decimal point in your final quotient, you need to compare the number of decimal places in the numerator with that of the denominator. There are three possibilities:

4(a) If the original numerator included *more* decimal places than did the denominator, shift the decimal point in your quotient to the *left* by the difference in the number of decimal places.

4(b) If the original numerator included *fewer* decimal places than did the denominator, shift the decimal point in your quotient to the *right* by the difference in that number of digits.

4(c) If the original numerator and denominator included the same number of decimal places, you're done! Don't shift the decimal point at all.

Here's how it works, using the division problem introduced earlier:

$$\frac{11.49}{.003}$$ original problem

$$\frac{1149}{3}$$ decimals and "leading" zero removed

383   stripped-down numbers combined by dividing

3830   decimal point shifted 1 place to the right in final quotient

Notice that we moved the decimal point one place to the right in the final step because the denominator of the original expression included one more decimal place than did the numerator (3 compared to 2). Had the situation been reversed, we would have moved the decimal point to the *left* instead:

$$\frac{1.149}{.03} = 38.3$$

Here, the numerator includes three decimal places, while the denominator includes only two. The additional one place in the numerator tells us to shift the decimal point in our quotient 383 one place to the *left*.

Had the numerator and denominator included the same number of decimal places, we would not have shifted the decimal point at all in our final quotient:

$$\frac{11.49}{.03} = 383$$

Percent

Percent means "per hundred" or "for every hundred." The term is derived from the Latin word *cent,* which means "one hundred." The symbol for percent is "%". (But you knew that, didn't you?) The term *percentage* is also used. (If you're wondering when to use "percentage" instead of "percent," consult an English grammar book—which this book isn't!)

One way to think about a percent is as a portion of a whole, where the whole is divided into 100 equal parts. Remember our pie from the fraction chapter? A whole pie can be divided into 100 equal slices; the number of slices you can eat is the percentage of the whole pie.

Percents can be greater than 100 as well. So if you eat more than 100% of a pie, there must have been at least one other pie on the dessert cart.

$$50\% = \text{half the pie}$$

$$100\% = \text{the whole pie}$$

$$200\% = \text{two whole pies}$$

Any percent can also be expressed as a fraction or as a decimal number. Here are some examples:

$$37\% = 37 \text{ out of } 100 = \frac{37}{100} = .37$$

$$284\% = \frac{284}{100} = 2.84$$

$$5\% = 5 \text{ out of } 100 = \frac{5}{100} = .05$$

$$3.4\% = 3.4 \text{ out of } 100 = \frac{3.4}{100} = \frac{34}{1,000} = .034$$

$$x\% = x \text{ out of } 100 = \frac{x}{100} = \left(\frac{1}{100}\right)(x) = .01x$$

$$\frac{1}{4}\% = \frac{1}{4} \text{ out of } 100 = \frac{\frac{1}{4}}{100} = \left(\frac{1}{100}\right)(.25) = .0025$$

A percent expressed as a *negative* number indicates *decrease*—from one number to a lower number. Otherwise, in our number system there is no such thing as a "negative percent." You can't bake a negative pie, can you?

Changing a Percent to a Decimal (and Vice Versa)

In the examples above, multiple steps were used to express percentages in terms of decimal equivalents. Here's a quicker way. To change a percent to a decimal, move the decimal point two places to the *left* and remove the percent sign. To change a decimal to a percent, move the decimal point two places to the *right* and add the percent sign. Here are three examples of each:

percent-to-decimal conversion	decimal-to-percent conversion
(shift decimal point two places to the *left*)	(shift decimal point two places to the *right*)
9.5% = .095	.003 = .3%
.4% = .004	.704 = 70.4%
123% = 1.23	13.661 = 1,366.1%

To multiply a decimal number by a percent, convert the percent to a decimal number, then multiply. If you wish to express the answer as a percent, move the decimal point back (to the right two places). Here's an example:

(.65)(43%)	original problem
(.65)(.43)	percent removed from second term (decimal point moved two places to the *left*)
.2795	two terms combined
27.95%	product expressed as a percentage (decimal point moved two places to the *right*)

If both original numbers are expressed as percents, don't convert either of them to decimal numbers. Just ignore all the percent signs and multiply one number by the other. Then put back the percent sign, moving the decimal point in the resulting product two places to the *left*. Here's an example:

12.8% of 70%	original problem
(12.8)(70)	percent removed from both terms

896   two terms combined

8.96%   product expressed as a percentage (decimal
                       point moved two places to the left)

Changing a Percent to a Fraction (and Vice Versa)

To change a percent to a fraction, simply remove the percent sign and divide by 100 (simplify by canceling common factors). Reverse the process to change a fraction to a percent (simplify by canceling common factors). Here is a potpourri of examples:

percent-to-fraction conversion	fraction-to-percent conversion
(divide by 100 and drop the "%")	(multiply by 100 and add a "%")

$$23\% = \frac{23}{100}$$

$$\frac{4}{5} = \left(\frac{4}{5}\right)(100\%) = \left(\frac{400}{5}\right)\% = 80\%$$

$$2.891\% = \frac{2.891}{100} \text{ or } \frac{2,891}{100,000}$$

$$\frac{3}{8} = \left(\frac{3}{8}\right)(100\%) = \left(\frac{300}{8}\right)\% = 37\frac{1}{2}\%$$

$$810\% = \frac{810}{100} = \frac{81}{10} \text{ or } 8\frac{1}{10}$$

Guarding Against Errors in Converting One Form to Another

To guard against conversion errors, keep in mind the general *size* of the number you are dealing with. If you're in a hurry, you might carelessly (and *incorrectly*) express...

$$.09\% \text{ as } .9 \text{ or as } \frac{9}{100} \text{ (wrong!)}$$

$$\frac{.4}{5} \text{ as } .8\% \text{ (wrong!)}$$

$$668\% \text{ as } 66.8 \text{ or as } .668 \text{ (wrong!)}$$

One good way to check your conversion is to verbalize the original expression, perhaps rounding off the original number to a more familiar one. In the first example above, think of .09% as just under .1%, which is one-tenth of a percent, or a thousandth (a pretty small number).

In the second example, think of $\frac{.4}{5}$ as just under $\frac{.5}{5}$, which is obviously $\frac{1}{10}$, or 10%. Think of 668% as more than 6 times a complete 100% or between 6 and 7. You'll be learning more about rounding off and other estimation techniques in Chapter 8.

Common Fraction/Decimal/Percent Equivalents

Certain fractions, decimals, and percents appear so commonly on math tests (and in life) that they're well worth memorizing. They are favorites of the testing services because they reward smart test-takers who recognize quicker ways of determining answers to questions. Learn your "fifths," sixths," and "eighths" until they are second nature to you. Here's a table to help you along:

fraction	*decimal*	*percent*
(increments of a "fifth")		
$\frac{1}{5}$	.2	20%
$\frac{2}{5}$	.4	40%
$\frac{3}{5}$	.6	60%
$\frac{4}{5}$	.8	80%

(increments of an "eighth")

$\frac{1}{8}$	.125	$12\frac{1}{2}\%$
$\frac{1}{4}$	.25	25%
$\frac{3}{8}$	.375	$37\frac{1}{2}\%$
$\frac{1}{2}$	.5	50%
$\frac{5}{8}$	.625	$62\frac{1}{2}\%$
$\frac{3}{4}$	.75	75%
$\frac{7}{8}$	.875	$87\frac{1}{2}\%$

(increments of a "sixth")

$\frac{1}{6}$	$.16\frac{2}{3}$	$16\frac{2}{3}\%$
$\frac{1}{3}$	$.33\frac{1}{3}$	$33\frac{1}{3}\%$
$\frac{1}{2}$	.5	50%
$\frac{2}{3}$	$.66\frac{2}{3}$	$66\frac{2}{3}\%$
$\frac{5}{6}$	$.83\frac{1}{3}$	$83\frac{1}{3}\%$

Computing Percentages

Exam problems involving percent typically combine percents and "raw" numbers (those not expressed as percents). They may require you to calculate a percent from a number or a number from a percent, or to add or subtract a percent from a number or percent. Though they might appear simple, these problems can easily lead to confusion. Let's sort out the different types and learn to handle each one in a smart way.

Finding a Percent of a Number

Consider this question: **What is 35% of 65?** There are three methods of finding a percent of a given number:

1. proportion method
2. decimal method
3. fractional method

Let's examine each one in turn.

1. Proportion method (set up an equation)

Equate 35%, or $\frac{35}{100}$, with the unknown number divided by 65. In doing so, you're asking, "35 is to 100 as what is to 65?" (Remember this technique from our look at "proportion" in Chapter 4?)

$$\frac{x}{65} = \frac{35}{100}$$

$$\frac{x}{65} = \frac{7}{20}$$

$$x = \frac{455}{20}$$

$$x = \frac{91}{4}$$

$$x = 22\frac{3}{4}$$

2. Decimal method (convert the percent to a decimal number)

Change 35% to .35, then multiply:

$$(.35)(65) = 22.75$$

3. Fractional method (convert the percent to a fraction)

Change 35% to $\dfrac{35}{100}$, then multiply:

$$\left(\frac{35}{100}\right)(65) = \left(\frac{7}{20}\right)(65) = \frac{(7)(13)}{4} = \frac{91}{4} = 22\frac{3}{4}$$

Most test-takers find the fractional method to be the easiest and quickest to work with, but you may prefer one of the other methods. In any event, always ask yourself whether the numbers given in a problem allow a quick conversion using the table on pages 76 and 77.

Finding a Number When a Percent Is Given

Consider the following question: **7 is 5% of what number?** The proportion method discussed above can also be used to find a number when a percent is given. However, most test-takers find it easier and more intuitive to set up a different equation—one that states algebraically exactly what the question asks verbally. Let's look at each method to answer the question.

1. Proportion method

$$\frac{5}{100} = \frac{7}{x} \quad \text{(You're asking: "5 is to 100 as 7 is to what?")}$$

$$5x = 700$$

$$x = 140$$

2. Verbal equation method

$$7 = .05x \quad \text{(You're asking: "7 is .05 of what?")}$$

$$700 = 5x$$

$$140 = x$$

Finding What Percent One Number Is of Another

Consider the following question: **12 is what percent of 72?** The proportion method can be used to find what percent 12 is of 72. However, you might find the fractional method easier and more intuitive. Let's apply each one to answer the question.

1. **Proportion method**

$$\frac{x}{100} = \frac{12}{72} \quad \text{(You're asking: "12 is to 72 as what is to 100?")}$$

$$x = \frac{1200}{72}$$

$$x = \frac{50}{3}$$

$$x = 16\tfrac{2}{3}\%$$

2. **Fractional method**

$$\frac{12}{72} = \frac{1}{6} = 16\tfrac{2}{3}\%$$

(You're asking: "What is the percent equivalent of 12/72?")

Percent of Increase or Decrease

Percent problems often involve percent *change* (increase or decrease). Consider this question: **10 increased by what percent is 12?** In handling questions such as this, keep in mind that the percent change always relates to the value *before* the change. First, determine the amount of the increase—2 in this case. Next, compare that increase to the original number (before the change) by a fraction: $\frac{2}{10}$ in this case. 10 increased by $\frac{2}{10}$ (or 20%) is 12.

The same procedure is used for percent decrease. Consider this question: **12 decreased by what percent is 10?** As before, first determine the amount of the change (2), then compare the change with the original amount (before the decrease):

12 in this case. The fractional decrease is $\frac{2}{12}$. 12 decreased by $\frac{2}{12}$ (or $\frac{1}{6}$ or $16\frac{2}{3}\%$) is 10. (Did you remember $16\frac{2}{3}\%$ from the conversion table on pages 76 and 77?)

Notice the percent increase from 10 to 12 (20%) differs from the percent decrease from 12 to 10 ($16\frac{2}{3}\%$). The reason for this is that the change is determined based on the original number (before the change), and that number is different in the first question and the second.

To underscore this important point, let's look at two more examples:

15 increased by 5 to 20 is a... 33.3% increase $\left(\frac{5}{15} \text{ or } \frac{1}{3}\right)$

but **20 decreased by 5 to 15 is a...** 25% decrease $\left(\frac{5}{20} \text{ or } \frac{1}{4}\right)$

25 increased by 25 to 50 is a... 100% increase $\left(\frac{25}{25} \text{ or } 1\right)$

but **50 decreased by 25 to 25 is a...** 50% decrease $\left(\frac{25}{50} \text{ or } \frac{1}{2}\right)$

Quiz Time

Here are 10 problems to test your skill in applying the concepts in this chapter. After attempting all 10 problems, read the explanations that follow. Then go back to the chapter and review your trouble spots. If you can handle the easier problems *and* the more challenging ones, consider yourself a "smart test-taker"!

Easier

1. Multiply, using the techniques for dealing with decimals discussed in this chapter:

   (a) $71.2 \times .6$

   (b) $32 \times .96$

   (c) 8.09×4.7

   (d) $.003 \times 23.07$

2. Divide, using the techniques for dealing with decimals discussed in this chapter:

   (a) $5.74 \div 1.4$

   (b) $71.2 \div .6$

   (c) $23.07 \div .003$

   (d) $.96 \div 32$

3. Convert fractions to percents, and convert percents to fractions:

   (a) $\frac{2}{12}$

   (b) $\frac{15}{24}$

   (c) 6%

   (d) 87.5%

4. Answer the following three questions: 45

   (a) What percent of 300 is 400?

   (b) What is 7% of 77?

   (c) 45 is 3,000% of what number?

5. A jewelry merchant paid $10,000 for a particular ring. For what price did the merchant later sell the ring, if the merchant's profit from the sale was 68.3%?

More Challenging

6. Express each of these numbers as a percent without performing long division. (Hint: These fractions are related to the common ones in the table on page 76.)

   (a) $\frac{5}{12}$

   (b) $\frac{7}{16}$

   (c) $\frac{33}{40}$

   (d) $\frac{16}{15}$

7. Answer the following three questions:

   (a) What percent of 80 is 65?

   (b) What is 111% of 17?

   (c) 225 is 180% of what number?

8. The temperature at 12:00 noon is 66 degrees. If the temperature increases by 25% during the afternoon, then decreases by 25% during the evening, what is the temperature at the end of the evening?

9. A clerk's salary is $320.00 after a 25% raise. Before the clerk's raise, the supervisor's salary was 50% greater than the clerk's salary. If the supervisor also receives a raise in the same dollar amount as the clerk's raise, what is the supervisor's salary after the raise?

10. Diane receives a base weekly salary of $800 plus a 5% commission on sales. In a week in which her sales totaled $8,000, what was the ratio of her total weekly earnings to her commission?

Answer and Explanations

1. Here are the answers to the four problems:

  (a) 42.72

  (b) 30.72

  (c) 38.023

  (d) .06921

2. Here are the answers to the four problems:

  (a) 4.1

  (b) $118\frac{2}{3}$

  (c) 7690

  (d) .03

3. Here are the answers to the four problems:

  (a) $16\frac{2}{3}$% (hint: $\frac{2}{12} = \frac{1}{6}$)

  (b) 62.5% (hint: $\frac{15}{24} = \frac{5}{8}$)

  (c) $\frac{3}{50}$

  (d) $\frac{7}{8}$

4. Here's how to approach the three questions:

  (a) The answer is: $133\frac{1}{3}$%. Think of 300 as 100%. 400 as $\frac{1}{3}$ ($33\frac{1}{3}$%) more than that 100%.

(b)   The answer is: 5.39. Multiply 7 by 77, then shift the decimal point in the re-sulting product, which is 539, to the left two places.

(c)   The answer is: 1.5. Set up the following equation: $45 = 30x$. Solving for x: $x = 45/30$ or 1.5.

5.   The answer is: $16,830. Given that the merchant paid $10,000 for the ring, if the merchant earned a 68.3% profit, the selling price of the ring was $10,000 plus 68.3% of $10,000.

6.   Here's how to quickly solve the four problems:

(a)   The answer is $41\frac{2}{3}\%$. Think of $\frac{5}{12}$ as midway between $\frac{4}{12}$ (which is $\frac{1}{3}$, or $33\frac{1}{3}\%$) and $\frac{6}{12}$ (which is $\frac{1}{2}$, or 50%). The midway point between $33\frac{1}{3}\%$ and 50% is $41\frac{2}{3}\%$.

(b)   The answer is $43\frac{3}{4}\%$. Think of $\frac{7}{16}$ as midway between $\frac{6}{16}$ (which is $\frac{3}{8}$, or $37\frac{1}{2}\%$) and $\frac{8}{16}$ (which is $\frac{1}{2}$, or 50%). The midway point between $37\frac{1}{2}\%$ and 50% is $43\frac{3}{4}\%$.

(c)   The answer is $82\frac{1}{2}\%$. Multiplying the denominator (40) by $2\frac{1}{2}$ gives you a denominator of 100. Accordingly, multiply 33 by $2\frac{1}{2}$ to obtain the answer.

(d)   The answer is $106\frac{2}{3}\%$. To change the denominator (15) to 100, divide by 3 and multiply the result by 20 ($15 \div 3 = 5 \times 20 = 100$). Accordingly, perform the same operations on the numerator ($16 \div 3 = 5\frac{1}{3} \times 20 = 106\frac{2}{3}$).

7.   Here's how to approach the three questions:

(a)   The answer is: $81\frac{1}{4}\%$. 65 is $\frac{65}{80}$ of 80. You can perform the division; or you can ask yourself: "65 is to 80 as what is to 100?" Set up an equation:

$$\frac{65}{80} = \frac{x}{100}$$

$$80x = (65)(100)$$

$$8x = 6500$$

$$x = \frac{6500}{8} \text{ or } 81.25$$

(b)   The answer is: 18.87. Either multiply 17 by 1.11 or add 11% of 17 to 17. Using the second method: $17 + (.11)(17) = 17 + 1.87 = 18.81$.

(c)   The answer is 125. Set up this equation: $225 = 1.8x$. Solve for x:

$$x = \frac{225}{1.8} \text{ or } 125$$

8.   The answer is $62\frac{3}{8}$ degrees. 25% of 66 is 16.5. Thus, the temperature increased to 82.5 degrees during the afternoon. 25% of 82.5 ($\frac{1}{4}$ of 82.5) is 20.625, or $20\frac{5}{8}$. Thus, the temperature decreased to $61\frac{7}{8}$ degrees during the evening.

9.   The answer is: $448. $320 is 125% of the clerk's former salary. Expressed algebraically:

$320 = 1.25x$

$32,000 = 125x$

$\$256 = x$ (clerk's salary before the raise)

Thus, the clerk received a raise of $64 ($320–$256). The supervisor's salary before the raise was:

$\$256 + 50\%$ of $\$256$
$\$256 + \$128 =$
$\$384$

The supervisor received a $64 raise. Thus, the supervisor's salary after the raise is $448 ($384 + $64).

10.   The answer is: 3:1. Diane's commission can be expressed as: $(.05)(8,000) = \$400$. Adding her commission to her base salary: $\$800 + \$400 = \$1,200$ (total earnings). The ratio of $1,200 to $400 is 3:1.

6

Exponents

Up to this point, we've avoided those tiny numbers known as *exponents* that hang above and to the right of regular numbers. We've also been avoiding *roots* and *radicals* up until now. Well, it's finally time to come to terms with these terms! Exponents and roots are two sides of the same coin, so to speak. To be more precise, they are two opposite ways of expressing "exponential" increase or decrease. In this chapter and the next, **you'll discover** that exponents and roots come with their own distinct "DO" and "DON'T" rules for combining them. **You'll also discover** that applying exponents or roots to numbers (or other terms) can yield interesting and often surprising results. Enough prologue, though. Let's get "exponential" and see what happens!

The Few Faces of Exponents

An *exponent* (or *power*) refers to the number of times that a number or other term (referred to as the *base*) is multiplied by itself. For example, $n^3 = (n)(n)(n)$. In this equation, n is said to be "raised to the third power" or "cubed." Here are four examples in which numbers are used as the base:

$$3^2 = (3)(3) = 9 \qquad\qquad 4^3 = (4)(4)(4) = 64$$

$$.2^4 = (.2)(.2)(.2)(.2) = .0016 \qquad \left(\frac{3}{5}\right)^3 = \left(\frac{3}{5}\right)\left(\frac{3}{5}\right)\left(\frac{3}{5}\right) = \frac{27}{125}$$

An exponent can be an integer or a fraction, a positive number or a negative number. Most exponents on standardized tests, however, are *positive integers*. Just for the record, however, here are examples and rules for negative and fractional exponents:

$$n^{-2} = \frac{1}{n^2}$$ raising a number to a negative exponent is the same as "1 over" the same number but with a positive exponent

$$n^{1/2} = \sqrt{n}$$

$$n^{2/3} = \sqrt[3]{n^2}$$ the numerator of a fractional exponent becomes the number's exponent; the denominator becomes the root

$$n^{3/2} = \sqrt{n^3}$$

For now, don't worry too much about the rules for negative and fractional exponents. We'll talk about them again later in this chapter.

The Exponential "Power" of Exponents

Statisticians often speak of particular numbers increasing "exponentially." This means that the numbers are increasing really fast as they are multiplied by themselves repeatedly. How fast? You're probably familiar with the classic illustration of this phenomenon, which we'll pose in the form of the following question:

> If I give you a penny today, and every day thereafter I give you double the total amount I've given you up to that point, how much money will I have given you altogether after 30 days?

> The answer is 2^{29} cents (or 2 multiplied by itself 29 times).

Why is the exponent 29 rather than 30? Because on the *second* day you'd have 2 pennies (2 cents). By the way, 2^{29} cents = \$536,870,912. Clearly, then, this is a hypothetical example, *not* a bonified offer on my part. We could have used any commodity to illustrate the point—rocks, for example. Unless you're a geologist, however, large amounts of money are more interesting than large numbers of rocks.

If my offer were to *triple* your money each day, the totals would be "exponentially" greater than those attained by doubling your money. And the later the date, the greater the exponential effect. For example, on the third day you'd receive 2 more pennies if your money is being doubled, and 6 more pennies if your money is being tripled (a difference of 4 pennies). Compare this to what happens on the fourth day—4 more pennies versus 18 more pennies, a difference of 14. So the difference between the totals increases with each passing day, as demonstrated here:

	Day 1	Day 2	Day 3	Day 4
double your money:	1	2	4	8
triple your money:	1	3	9	27

In fact, it would take only 19 days of tripling your money each day to exceed the amount of money you would have after 30 days of doubling your money each day. By the way, after tripling your money daily for 30 days, you'd have 3^{29} dollars, or $68,630,377,364,833. That's almost 69 trillion dollars.

Exponential Versus Linear Relationships

Each of the get-rich-quick schemes we just described is an example of an "exponential relationship"; the relationship is between the number of days and the number of pennies). On each successive day, not only did your total increase, but the amount of the increase increased. How does this differ from a "linear relationship"? Well, if I instead gave you *the same* additional number of pennies on each successive day, the relationship between the number of days and the number of pennies would remain proportionately the same. To understand why this relationship would be called "linear," let's graph both types of relationships. We'll use a vertical axis to indicate total pennies, and we'll use the horizontal axis to indicate days. Plotting points for days 1–6 and connecting the points, here's what we get:

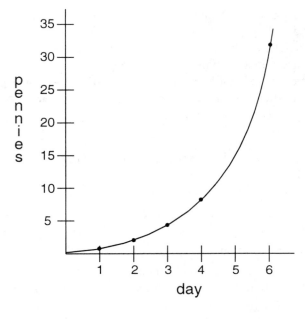

double your money daily
(exponential relationship)

day	total
1	1
2	2
3	4
4	8
5	16
6	32

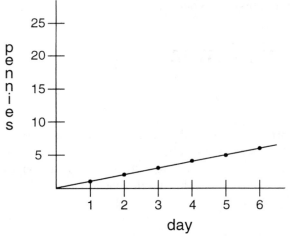

add 1 penny each day
(linear relationship)

day	total
1	1
2	2
3	3
4	4
5	5
6	6

Notice that the linear relationship is a straight *line*, while the "exponential relation-ship" is a curve which extends up much further (much faster) than it extends to the right. We could go further and express each relationship as an algebraic equation, but we would be getting *way* ahead of ourselves.

Exponents and the Real Number Line

In the preceding section, we saw how numbers can *increase* exponentially when raised to powers. This isn't always the case, though. Numbers can also decrease exponentially. Under what circumstances would this happen? Well, let's think about non-integers between 0 and 1 on the number line. These numbers will *decrease* (at an decreasing rate) as you raise them to a power that is a positive integer:

$$\left(\frac{1}{2}\right)^2 = \left(\frac{1}{2}\right)\left(\frac{1}{2}\right) = \frac{1}{4}$$ result is 1/2 the size of the original number

$$\left(\frac{1}{2}\right)^3 = \left(\frac{1}{2}\right)\left(\frac{1}{2}\right)\left(\frac{1}{2}\right) = \frac{1}{8}$$ result is 1/4 the size of the preceding number

$$\left(\frac{1}{2}\right)^4 = \left(\frac{1}{2}\right)\left(\frac{1}{2}\right)\left(\frac{1}{2}\right)\left(\frac{1}{2}\right) = \frac{1}{16}$$ result is 1/8 the size of the preceding number

If we kept going, multiplying each resulting fraction by 1/2, we would continue indefinitely to cut our number in half. Notice that each incremental decrease is *smaller* (not larger) than the previous one—our number is decreasing at a *decreasing* rate as we increase the value of the exponent. Also notice that our incredible shrinking number would approach, but never reach, zero. By the way, we'd see the same pattern with larger exponents (such as 3 or 4); the numbers would just get smaller faster.

What if we instead used a non-integer *between 0 and –1* as our base number? What if we used a number *less than –1* (such as –2) as our base number? You'd better know the answers to these questions if you're going to consider yourself a smart test-taker. This is a favorite area in standardized testing! Let's organize our observations about the impact of exponents on the size of a number. In doing so, try to visualize these numbers on the real number line, and strive to "internalize" these patterns so that they are second nature to you.

General Pattern	Examples:
A *positive* base number... raised to *any exponent...* results in a *positive* number.	$\left(\dfrac{1}{3}\right)^3 = \dfrac{1}{27}$, $9^{-2} = \dfrac{1}{81}$, $5^{1/2} \approx 2.2$
A *negative* base number... raised to an *even integer...* results in a *positive* number.	$-4^2 = (-4)(-4) = +16$ $-3^4 = (-3)(-3)(-3)(-3) = +81$
A *negative* base number... raised to an *odd integer...* results in a *negative* number.	$\left(-\dfrac{1}{2}\right)^3 = \left(-\dfrac{1}{2}\right)\left(-\dfrac{1}{2}\right)\left(-\dfrac{1}{2}\right) = -\dfrac{1}{8}$ $-2^5 = (-2)(-2)(-2)(-2)(-2) = -32$
A base number greater than 1... raised to an exponent greater than 1... results in a larger number.	$(1.1)^2 = (1.1)(1.1) = 1.21$ $2^{3/2} = \sqrt{2^3} \approx 2.8$
A base number between 0 and 1... raised to an exponent greater than 1... results in a smaller number approaching 0.	$\left(\dfrac{1}{3}\right)^2 = \left(\dfrac{1}{3}\right)\left(\dfrac{1}{3}\right) = \dfrac{1}{9}$ $.6^3 = (.6)(.6)(.6) = .216$
A base number between 0 and −1... raised to an odd exponent greater than 1... results in a larger number (to the right on the number line) approaching 0.	$-.4^3 = (-.4)(-.4)(-.4) = -.64$ $-.2^5 = (-.2)(-.2)(-.2)(-.2)(-.2)$ $= -.00032$

A base number *less than −1* (left of −1 on the number line)... raised to an *odd* exponent *greater than 1*... results in a *smaller* number (to the left on the number line).

$$-\left(\frac{3}{2}\right)^3 = \left(-\frac{3}{2}\right)\left(-\frac{3}{2}\right)\left(-\frac{3}{2}\right) = -\frac{27}{8}$$

$$-5^5 = (-5)(-5)(-5)(-5)(-5) = -3{,}125$$

What about the impact of *negative* exponents on base numbers? Well, a negative exponent has just the opposite effect on a base number that a positive exponent has. Here are two examples:

$$3^{-2} = \frac{1}{3^2} \text{ or } \frac{1}{9}$$

negative integral exponent *decreases* the value of a positive number greater than 1

$$\left(\frac{2}{3}\right)^{-2} = \frac{1}{\left(\frac{2}{3}\right)^2} = \left(\frac{3}{2}\right)^2 \text{ or } \frac{9}{4}$$

negative integral exponent *increases* the value of a non-integer between 0 and 1

Go ahead and try applying negative exponents to negative base numbers. In each case, you'll see that they have an impact that is opposite to that of positive numbers—decreasing instead of increasing (and vice versa) the value of a base number.

As for *fractional exponents* and their impact on base numbers, fractional exponents are actually roots in disguise, so we'll talk about them in Chapter 7.

Remember: The impact of raising a number to an exponent (power) depends on the region on the number line where the number and exponent fall. The four regions are:

1. less than (to the left of) −1

2. between −1 and 0

3. between 0 and 1

4. greater than (to the right of) 1

DO's and DON'Ts for Combining Base Numbers and Exponents

Can you combine base numbers—using addition, subtraction, multiplication, or division—*before* applying exponents to the numbers? That depends on which operation you're performing. Here's a good place to start:

addition or subtraction: RED light!! Don't combine base
                                 numbers (or exponents)

multiplication or division: YELLOW light! Combine base
                                 numbers or exponents, but only
                                 under certain conditions

Now let's see how these traffic rules play out "on the street."

Addition and Subtraction

When you add or subtract terms, you cannot combine base numbers or exponents. It's as simple as that. Let's express this prohibition in terms of the intrepid variables a, b, and x:

$$a^x + b^x \neq (a+b)^x$$

Take a look at a simple example. Notice that you get a different result depending on which you do first: combine base numbers or apply each exponent to its base number. This is a huge clue that something's wrong!

$$3^2 + 4^2 = (3+4)^2 = 7^2 = 49 \quad \text{base numbers combined first (improper)}$$

$$3^2 + 4^2 = 9 + 16 = 25 \quad \text{exponents applied first (proper)}$$

Multiplication and Division

It's a whole different story for multiplication and division. First, remember these two simple rules:

1. You can combine base numbers first, but only if the exponent is the same.

2. You can combine exponents first, but only if the base number is the same.

Let's assume first that exponents are the same:

general rule:

$$a^x \cdot b^x = (ab)^x$$

example:

$$2^3 \times 3^3 = (2 \times 3)^3 = 6^3 = 216$$
$$2^3 \times 3^3 = (8)(27) = 216$$

$$\frac{a^x}{b^x} = \left(\frac{a}{b}\right)^x$$

$$\frac{2^3}{3^3} = \left(\frac{2}{3}\right)^3 = \frac{2}{3} \cdot \frac{2}{3} \cdot \frac{2}{3} = \frac{8}{27}$$

Now let's assume that base numbers are the same. When multiplying such terms, add the exponents. When dividing them, subtract the denominator exponent from the numerator exponent:

general rule:

$$a^x \cdot a^y = a^{x+y}$$

examples:

$$2^3 \cdot 2^2 = 2^{(3+2)} = 2^5 \text{ or } 32$$

$$\frac{a^x}{a^y} = a^{x-y}$$

$$\frac{2^5}{2^2} = 2^{(5-2)} = 2^3 \text{ or } 8$$

$$\frac{2^2}{2^5} = 2^{(2-5)} = 2^{-3} = \frac{1}{2^3} = \frac{1}{8}$$

Remember: You can't combine base numbers or exponents of two terms unless you are multiplying or dividing the terms *and* either base numbers are the same or exponents are the same.

Additional Exponent Rules

Here are a few more exponent rules to round out the discussion. The first one is covered far more commonly on standardized exams than the others.

general rule	example

$$(a^x)^y = a^{xy} \qquad\qquad (2^2)^4 = 2^8 = 256$$

$$a^{x/y} = \sqrt[y]{a^x} \qquad\qquad 4^{2/3} = \sqrt[3]{4^2} = \sqrt[3]{16} \approx 2.5$$

$$a^{-x} = \frac{1}{a^x} \qquad\qquad 3^{-2} = \frac{1}{3^2} = \frac{1}{9}$$

$$a^{-x/y} = \frac{1}{a^{x/y}} \qquad\qquad 3^{-3/2} = \frac{1}{3^{3/2}} = \frac{1}{\sqrt{3^3}} = \frac{1}{\sqrt{27}} = \frac{1}{3\sqrt{3}} = \frac{\sqrt{3}}{9}$$

$$a^0 = \quad (a \neq 0) \qquad\qquad 34^0 = 1$$

Exponential Values You Should Memorize

The exponential values (numbers) indicated in the following table appear so frequently on standardized exams that it is worthwhile to commit them to memory.

Base	Power and Corresponding Value						
	2	3	4	5	6	7	8
2	4	8	16	32	64	128	256
3	9	27	81	243			
4	16	64	256				
5	25	125	625				
6	36	216					

z^2 4

z^3 16

Quiz Time

Here are 10 problems to test your skill in applying the concepts in this chapter. After attempting all 10 problems, read the explanations that follow. Then go back to the chapter and review your trouble spots. If you can handle the easier problems *and* the more challenging ones, consider yourself a "smart test-taker"!

Easier

1. If $-1 < x < 0$, rank the values of the following five terms, from largest to smallest:

$-\frac{1}{2}$

(a) x^2 (b) x^3 (c) x^0 (d) $-x$ (e) $\dfrac{1}{x^3}$

2. Express each of the following as a decimal number:

(a) 3.2×10^{-2} (b) $.123 \times 10^4$ (c) 7.007×10^{-3}

(d) 9.1×10^0 (e) 1.44×10^6

3. If $x = -1$, find the value of each these expressions:

(a) $x^4 - x^3 - x^2 - x^1 - x^0$

(b) $\dfrac{x^5}{x^7} + \dfrac{x^7}{x^5}$

(c) $\dfrac{x^3}{x^2} - \dfrac{x^2}{x^3}$

(d) $\dfrac{x^{38} x^{37}}{x^{38} - x^{37}}$

(e) $x^{-3} + x^{-2} + x^2 + x^3$

4. Determine the value of each of these expressions using the techniques in this chapter, without resort to a calculator or to pencil and paper:

(a) $\dfrac{4^4}{64}$ (b) $\dfrac{2^6}{56}$ (c) $\dfrac{27}{3^5}$ (d) $5^3 - 2^7$ (e) $2^8 - 4^4$

5. Determine the value of each of these expressions using the techniques in this chapter, without resort to a calculator or to pencil and paper:

(a) $\dfrac{27^{27}}{27^{26}}$ (b) $.5^5 \cdot 4^5$ (c) $\dfrac{3^3 \cdot 3^4 \cdot 3^5}{3^{15}}$

(d) $\dfrac{6^3}{3^3}$ (e) $\left(-2^{-2}\right)^{-3}$ (f) $\dfrac{9^{11} \cdot 11^9}{11^9 \cdot 9^9}$

More Challenging

6. Assume that $|x| < 1$ and that $x \neq 0$. Rank the following five terms, from largest to smallest in value:

(a) x^4 (b) $\dfrac{1}{x^2}$ (c) $-\left|\dfrac{1}{x^3}\right|$ (d) $-(x^5)$ (e) $-\dfrac{1}{x^2}$

7. $7^{77} - 7^{76} =$

(A) 7   (B) $7^{77/76}$ (C) 49   (D) $6 \cdot 7^{76}$ (E) 7^{75}

8. In each of these four expressions, what is the value of x, assuming the value of the expression is zero? (Analyze the problems intuitively, applying the concepts in this chapter rather than solving the problems algebraically.)

(a) $x^2 - 2x$

(b) $x^2 - x^3$ (there are two possible values of x)

(c) $x + x^2$

(d) $\dfrac{1}{x} - x^3$

9. If $\dfrac{x}{y}$ is a negative integer, which of the following terms must also be a negative integer?

(a) $\dfrac{x^2}{y}$ (b) $-\dfrac{x^2}{y^2}$ (c) $\dfrac{x}{y^2}$ (d) $x + y$ (e) xy

10. Consider (a) through (d) individually. Is the information provided sufficient to determine whether xyz is a *positive* number?

(a) $x^3 y^2 z < 0$ (b) $xyz^3 < 0$ (c) $xy^2 z > 0$ (d) $x^2 y^4 z^2 > 0$

Answers and Explanations

1. From largest to smallest, the order is: (c), (d), (a), (b), (e).

(c) equals 1

(d) is a positive number between 0 and 1.

(a) is a positive number between 0 and |x|, which is the value of (d)

(b) is a positive non-integer between 0 and x^2, which is the value of (a)

(e) is a negative number less than (to the left of) -1

2. Here are the answers to the five problems:

  (a) .032

  (b) 1,230

  (c) .007007

  (d) 9.1

  (e) 1,440,000

3. Here are the answers to the four problems:

  (a) 2

  (b) 2

  (c) 0

  (d) $-\frac{1}{2}$

  (e) 0

4. Here are the answers to the six problems:

  (a) 4

  (b) $\frac{8}{7}$

  (c) $\frac{1}{9}$

  (d) −3

  (e) 0

5. Here are the answers to the six problems:

  (a) 27   (b) 32   (c) $\frac{1}{27}$ (d) 8   (e) 64   (f) 81

6. From largest to smallest, the order is: (b), (a), (d), (e), (c). Given that $|x| < 1$ and that $x \neq 0$, x must be a non-integer greater than −1 but less than 1.

  (b) must be greater than 1.

  (a) must be a positive number between 0 and 1.

(d) could be either a negative number between −1 and 0 or positive number between 0 and 1. In either case, however, (c) must lie to the *left* of (d) on the number line.

(e) must be a negative number less than (to the left of) −1.

(c) must be a negative number less than (to the left of) (e) on the number line.

7. The correct answer is (D). The expression involves subtraction, neither the base numbers nor the exponents can be combined. Only (D) is equivalent to the original expression. To confirm this without a calculator, try using smaller numbers that exhibit the same pattern; for example:

$3^4 - 3^3 = 81 - 27 = 54$, which is equivalent to $2 \cdot 3^3$.

8. Here are the answers to the four problems:

(a) $x = 0$ or 2. Factor out an x from each term, then set both roots equal to 0:

$x(x - 2) = 0$

$x = 0, \ x - 2 = 0$

(b) $x = 0$ or 1. x^2 must equal x^3 if their difference is 0. The only case in which these two terms could be equal is if $x =$ either 1 or 0.

(c) $x = -1$ or 0. Given that $x + x^2 = 0$, $x(1 + x) = 0$.

(d) $x = 1$ or −1. The two terms must equal each other. The only values of x that will satisfy this condition are 1 and −1.

9. Of the five expressions, only (b) *must* be a negative integer, even if x and y are not themselves integers. Because the overall fraction is an integer, $\dfrac{x^2}{y^2}$ must be an integer. Any number squared is positive, so $\dfrac{x^2}{y^2}$ must be positive. Accordingly, $-\dfrac{x^2}{y^2}$ must be negative.

(a) can be either a positive or negative integer, depending on whether y is positive or negative.

(c)   can be a non-integer, since the denominator of the original expression is squared. Also, (c) can be either positive or negative, depending on the sign of x.

(d)   can must be an integer, but can be either positive or negative, depending on whether x is negative or y is negative.

(e)   must be negative, but it is not necessarily an integer—for example:

$$\frac{-\frac{2}{3}}{\frac{2}{3}} = -1, \text{ but } -\frac{2}{3} \cdot \frac{2}{3} = -\frac{4}{9} \text{ (a non-integer)}$$

10. In all four cases, note that neither x, y, nor z can equal zero (otherwise, their product would be zero). Here is a specific analysis of each problem:

(a)   is not sufficient to determine whether xyz is negative. Given that $x^3 y^2 z < 0$, either all three terms (x^3, y^2 and z) are negative or exactly one of the three terms is negative. However, whether y is negative or positive, y^2 must positive; thus either x or z (but not both) must be negative. Accordingly, xyz could be either positive or negative, depending on the value of y.

(b)   is sufficient to answer the question. Either z, y, and x are all negative or exactly one of the three variables is negative. In either case, $xyz < 0$.

(c)   is insufficient to answer the question. Given that $xy^2 z > 0$, x and z must both be either negative or positive, but y can be either negative or positive. Accordingly, xyz could be either negative or positive, depending on y's value.

(d)   is insufficient to answer the question. Any variable, positive or negative, raised to an even exponent will be even. Thus, x, y, and z could each be either negative or positive, given that $x^2 y^4 z^2 > 0$.

7

Roots and Radicals

On the flip side of the exponential world is the realm of roots and radicals. What is a "root" in mathematics? Well, consider the roots of a tree, which are hidden beneath the surface and from which the tree grows. In mathematics the "root" of a given number is some other hidden number, from which the visible number has grown. Let's take this analogy one step further. The nutrients used to grow the hidden root into the visible number (the tree) are exponents!

Square roots and *cube* roots are the two most common root varieties. Here's a good definition of each one:

The **square root** of a number is another number that multiplied by itself equals the first number.

The **cube root** of a number is another number that multiplied twice by itself equals the first number.

Roots are denoted by "radical" signs ($\sqrt{}$). The term within the sign is the "tree"—the number whose root you need to determine. The smaller number outside and to the left of the sign indicates which root you need to dig for. If no number is given, then it is the *square* root you must determine. Let's look at a few examples to see how roots are related to exponents:

square root of 4	$\sqrt{4} = 2$	$2^2 = (2)(2) = 4$
cube root of 27	$\sqrt[3]{27} = 3$	$3^3 = (3)(3)(3) = 27$
fourth root of .0625	$\sqrt[4]{.0625} = .5$	$.5^4 = (.5)(.5)(.5)(.5) = .0625$

On standardized tests, you will not be expected to calculate exact roots (square or otherwise) of large numbers. You may, however, be asked to *approximate* or determine a range of values of square roots (or other roots). For example, $\sqrt[3]{50}$ lies somewhere between 3 and 4 ($3^3 = 27$, $4^3 = 64$). This range would probably be sufficient to respond to the question at hand.

Roots and the Real Number Line

As with exponents, the root of a number can bear a surprising relationship to the *size* and/or *sign* (negative vs. positive) of the number (another favorite area of the testing services). Here are four observations you should remember:

Observation #1:
If $n > 1$, then $1 < \sqrt[3]{n} < \sqrt{n} < n$ (the higher the root, the lower the value). However, if n lies between 0 and 1, then $n < \sqrt{n} < \sqrt[3]{n} < 1$ (the higher the root, the higher the value).

(example: $n = 64$)   (example: $n = \frac{1}{64}$)

$1 < \sqrt[3]{64} < \sqrt{64} < 64$ $\frac{1}{64} < \sqrt{\frac{1}{64}} < \sqrt[3]{\frac{1}{64}} < 1$

is the same as   *is the same as*

$1 < 4 < 8 < 64$ $\frac{1}{64} < \frac{1}{8} < \frac{1}{4} < 1$

Observation #2:

The *square* root of any negative number is an *imaginary* number, not a real number. Remember: you won't encounter imaginary numbers on basic math tests.

Observation #3:

Every negative number has exactly one cube root, and that root is a negative number $[(-)(-)(-) = (-)]$. The same holds true for all other *odd*-numbered roots of negative numbers. Here are two examples:

$$\sqrt[3]{-27} = -3 \qquad (-3)(-3)(-3) = -27$$

$$\sqrt[5]{-32} = -2 \qquad (-2)(-2)(-2)(-2)(-2) = -32$$

Observation #4:

Every positive number has two square roots: a negative number and a positive number (with the same absolute value). The same holds true for all other even-numbered roots of positive numbers. Here are two examples:

$$\sqrt{16} = +4 \text{ or } -4 \qquad\qquad \sqrt[4]{81} = +3 \text{ or } -3$$

However, every positive number has only one cube root, and that root is always a positive number. The same holds true for all other odd-numbered roots.

DOs and DON'Ts for Combining Roots

The rules relating to combining terms that include roots are quite similar to those for exponents. Keep two rules in mind; one applies to addition and subtraction, while the other applies to multiplication and division:

1. **Addition and subtraction.** If a term under a radical is being added to or subtracted from a term under a different radical, you cannot combine the two terms under the same radical.

general rule	examples
$\sqrt{x} + \sqrt{y} \neq \sqrt{x+y}$	$\sqrt{4} + \sqrt{16} = 2+4 = 6$ *but...*
	$\sqrt{4+16} = \sqrt{20} \approx 4.4$
$\sqrt{x} - \sqrt{y} \neq \sqrt{x-y}$	$\sqrt{25} - \sqrt{9} = 5-3 = 2$ *but...*
	$\sqrt{25-9} = \sqrt{16} = 4$
$\sqrt{x} + \sqrt{x} = 2\sqrt{x}$	$\sqrt{36} + \sqrt{36} = 2\sqrt{36} = 2(6) = 12$ *but...*
(not $\sqrt{2x}$ *)*	$\sqrt{2 \cdot 36} = \sqrt{72} \approx 8.5$

2. **Multiplication and Division.** Terms under different radicals *can* be combined under a common radical if one term is multiplied or divided by the other, *but only if* the root is the same. Here are three different cases:

general rule	example
$\sqrt{x}\sqrt{x} = x$	$\sqrt{33.9}\sqrt{33.9} = 33.9$
$\sqrt{x}\sqrt{y} = \sqrt{xy}$	$\sqrt{9}\sqrt{4} = \sqrt{9 \cdot 4} = \sqrt{36} = 6$
$\dfrac{\sqrt{x}}{\sqrt{y}} = \sqrt{\dfrac{x}{y}}$	$\dfrac{\sqrt[3]{125}}{\sqrt[3]{8}} = \sqrt[3]{\dfrac{125}{8}} = \dfrac{5}{2}$

Simplifying Radicals

There are numerous ways to "simplify" terms that include radicals. You might be able to get rid of a radical sign altogether; or perhaps you can move part of what's inside the radical to the outside. Let's look at these and a few other possibilities.

Eliminating Radicals Altogether

When and how do you get rid of radicals altogether? First, check to see if the root value is a specific value that you already know. For example, we can just get rid of these radicals altogether…no fuss, no muss:

$$\sqrt{25} = 5$$

$$\sqrt[3]{8} = 2$$

$$\sqrt[4]{a^4} = a$$

Does this mean you need to memorize extensive root tables for your exam? No. All the ones you should know for quick reference are listed later in this chapter, and there aren't very many to learn.

You can also get rid of some radical signs that appear in *equations*. If the root is a square root, for example, just square both sides of the equation to get rid of the radical. Here's an example:

$$\sqrt{x + 2y} = 7$$

$$\left(\sqrt{x + 2y}\right)^2 = (7)^2$$

$$x + 2y = 49$$

If this equation had involved a *cube* root instead, we could have "cubed" both sides of the equation to rid ourselves of the pesky radical:

$$\sqrt[3]{x + 2y} = 7$$

$$\left(\sqrt[3]{x + 2y}\right)^3 = (7)^3$$

$$x + 2y = 343$$

Moving Part of What's Inside a Radical to the Outside

If you can't get rid of a radical sign altogether, you might be able to move part of what's inside the radical to the outside. Check inside your square-root radicals for factors that are squares of nice tidy numbers (especially integers). Remove the term from the radical, and place its root value preceding the radical to signify multiplication. You might also be able to remove variables that are raised to other powers from the shackles of radicals. Here's an example that includes both a number and a variable:

$$\sqrt{\frac{20}{3}x^3} = \sqrt{\frac{(4)(5)}{3}(x^2)(x)} = 2x\sqrt{\frac{5}{3}x}$$

Notice that we removed 4 (a factor of 20) and an x^2 from within the radical, placing their square roots (2 and x) outside instead. This advice applies to other roots as well; for example:

$$\sqrt[3]{\frac{3}{8}} = \frac{1}{2}\sqrt[3]{3} \text{ or } \frac{\sqrt[3]{3}}{2}$$

Moving a Radical from a Denominator to a Numerator

If you can't get rid of a radical sign, you might still be able to push it around a bit. Let's say one of those pesky radical signs shows up in a denominator of a fraction. In math etiquette, leaving a radical in a denominator is a bit of a *faux pas* (pardon my French). In fact, be assured that on standardized math tests, the "correct" answer won't include a radical in a denominator. To remove a radical from a denominator, multiply both numerator and denominator by the radical value. Here's an example:

$$\frac{3}{\sqrt{15}} = \left(\frac{\sqrt{15}}{\sqrt{15}}\right)\frac{3}{\sqrt{15}} = \frac{3\sqrt{15}}{15} \text{ or } \frac{\sqrt{15}}{5}$$

Remember: On your exam, if you can reduce an expression under a radical sign by removing terms or factors, do it! Also, eliminate radicals from denominators. More than likely, these steps will be necessary to solve the problem at hand.

Expressing a Radical as an Exponent Instead

You can also get rid of a radical sign by expressing it as an exponent instead. This doesn't exactly simplify matters, however. You're just jumping out of a frying pan into a fire. In fact, in most cases you're actually complicating things by using an exponent instead of a radical:

general rule

$$\sqrt[x]{a^y} = a^{y/x}$$

examples

$$8^{1/3} = \sqrt[3]{8} = 2$$

$$\sqrt{7^3} = 7^{3/2} = 343^{1/2}$$

As noted in Chapter 6, fractional exponents do not appear frequently on standardized exams.

Roots and Decimal Places

Do you panic when you see decimal points in a "root" problem? If so, it's time to calm down and figure out how to handle those pesky little dots. Let's consider the following square root (one that you definitely should memorize):

$$\sqrt{144} = 12$$

What happens to the square root if we move the decimal point in the number 144 to the left or right? It depends on whether you move it an *odd* number of places (1,3,...) or an *even* number of places (2,4,...). What pattern can you discern from the following table? (The equations in the right-hand column might shed some light on this.)

This amount:	is the square root of a number that shifts the decimal point in 144:
$\sqrt{.0144} = .12$	(4 places to the left)
$\sqrt{.144} \approx .3795$	(3 places to the left)
$\sqrt{1.44} = 1.2$	(2 places to the left)
$\sqrt{14.4} \approx 3.795$	(1 place to the left)
$\sqrt{144} = 12$	(0 places)
$\sqrt{1,440} \approx 37.95$	(1 place to the right)
$\sqrt{14,400} = 120$	(2 places to the right)
$\sqrt{144,000} \approx 379.5$	(3 places to the right)
$\sqrt{1,440,000} = 1,200$	(4 places to the right)

Here's what's happening. When you move the decimal point an even number of places (either left or right), the root is the same, but with the decimal point moved *half as many places* in the same direction. However, if you move the decimal point an *odd* number of places (left or right), your root becomes an entirely different number. This makes perfect sense if you think about it terms of smaller numbers—for example:

$$\sqrt{9} = 3$$

$$\sqrt{90} = \text{ a number about midway between 9 and 10}$$

$$\sqrt{900} = 30$$

Try it with a few others such as $\sqrt{16}$ and $\sqrt{25}$. Just remember to think twice when you see one or more zeros following a number under a radical sign. It could be one of the test-makers' favorite traps, so don't fall for it!

Determining Square-Root and Cube-Root Values

Aside from using a calculator or asking a mathematician, there are three ways to determine roots of specific numbers:

1. memorize root tables

2. learn how to calculate precise root values

3. learn how to approximate root values

For standardized exams, don't bother with the first two approaches. You will not be expected to determine precise root values, except for certain "nice round" numbers. So we're not going to provide extensive root tables in this book, nor will we will take the time to show you how to calculate precise roots (it's time-consuming and tricky). On standardized exams, aside from a handful of "common" roots, approximations suffice.

So what about those "common" roots? Well, you've already seen a table of common roots. Take another look at the table of common exponents on page 97 in Chapter 6. Just read this table "the other way," and you'll learn common roots! (Remember: roots are just the flip side of exponents.) Here's another table for you to help memorize common square roots and cube roots:

common square roots	common cube roots
$\sqrt{121} = 11$	$\sqrt[3]{8} = 2$
$\sqrt{144} = 12$	$\sqrt[3]{27} = 3$
$\sqrt{169} = 13$	$\sqrt[3]{64} = 4$
$\sqrt{196} = 14$	$\sqrt[3]{125} = 5$
$\sqrt{225} = 15$	$\sqrt[3]{216} = 6$
$\sqrt{625} = 25$	$\sqrt[3]{343} = 7$
	$\sqrt[3]{512} = 8$
	$\sqrt[3]{729} = 9$
	$\sqrt[3]{1000} = 10$

$\sqrt{16} = 4$

$\sqrt{160}$

$\sqrt{1600} = 40$

Approximating Square Roots by Interpolating from Known Values

Square roots of many two- and three-digit numbers can be approximated by *interpolating* from the common roots listed in the table above. Interpolation involves finding two values between which the root in question lies.

Let's start with an easy one. We'll approximate the value of $\sqrt{163}$ by interpolating from the numbers provided in our root table. Ask yourself:

> **"What two consecutive integers can I square so that the square of the lower integer is less than 163 while the square of the greater integer exceeds 163?"**

According to the root table, those two integers are 12 and 13:

$$12^2 = 144 \qquad\qquad 13^2 = 169$$

163 lies between 144 and 169, and so $\sqrt{163}$ must lie between 12 and 13. Is this approximation close enough? Well, it depends on how precise a value the particular exam question requires. If you need a narrower range, you can add this extra step:

> **Since 163 is closer to 169 than to 144,**
> $\sqrt{163}$ **should be closer to 13 than to 12.**

Indeed, it is. Just for the record, $\sqrt{163} = 12.76$ (to the nearest hundredth). But the estimate of "more than 12.5 but less than 13" should suffice. Be careful with this extra step! *Don't* get the idea that if the square of a number lies exactly halfway (or one-third of the way) from one integer to the next, then it's square root must lie *exactly* halfway (or one-third of the way) from the square root of the lower number to that of the higher one. For example:

> **127.5 is midway between 121 and 144.**
> $\sqrt{121} = 11$, **and** $\sqrt{144} = 12$.

Does this mean that $\sqrt{127.5}$ equals 11.5 (midway between 11 and 12)? No! $\sqrt{127.5} \approx 11.29$ (closer to 11 than to 12). The reason for this is that when we take the square root of a number, we are not dividing it by two, just as we are not multiplying a number by 2 when we square it. One is a linear operation; the other is an exponential one.

Approximating Square Roots by Grouping Digits

Okay, as long as the numbers involved are short enough (two and three digits) you can easily interpolate from roots that you already know to estimate another root. But what about longer numbers—with four or five digits (or more)? (I use the word "long" because a long number is not necessarily big; it could include many places to the right of a decimal point.) Except for those "nice round numbers" like the common roots listed on page 111, you won't be asked on basic math exams to determine *precise* values of roots for long numbers. However, an exam question may require that you *approximate* them. Here's a quick way to estimate square roots of long numbers. Let's use the number $\sqrt{7,562}$:

1. Separate the number's digits into pairs: (75) (62). (Always start from the right, so that any unpaired digit appears at the far left.)

2. Starting with the pair to the far left (or unpaired digit at the far left), ask yourself: "What integer comes closest to the square root of this number (75) without exceeding it? $\sqrt{75}$ is somewhere between 8 and 9, so the answer is 8.

3. The first digit of $\sqrt{7,562}$ is 8.

4. For each digit pair to the right of (75), add *one* zero to "8": 80.

5. $\sqrt{7,562}$ is more than 80 but less than 90. This probably will suffice to answer your exam question.

If necessary, you can always narrow the range further:

1. $80^2 = 6,400$

2. $90^2 = 8,100$

3. 7,562 is closer to 8,100 than to 6,400

4. Thus, $\sqrt{7,562}$ is closer to 90 than to 80

By the way, $\sqrt{7,562} \approx 86.96$. Let's try one more, this time using a number with an *odd* number of digits:

$\sqrt{58,339}$	original number
5(83)(39)	pairing the digits (from right to left)
the first digit is 2	$\sqrt{5}$ is between 2 and 3
$\sqrt{58,339}$ is between 200 and 300	add a zero for every digit pair to the right of "5"
$\sqrt{58,339}$ is between 200 and 250	58,339 is closer to 200^2 than to 300^2
242	$\sqrt{58,339}$ (to the nearest unit)

Approximating Cube Roots by Grouping Digits

You can estimate cube roots using the same technique you did for estimating square roots, with some slight modifications. Instead of pairing the digits, separate them into *groups of three*. Also, instead of estimating the square root of the first group, estimate the *cube* root of that group. Let's use 58,339 again to see how it works:

$\sqrt[3]{58,339}$	original number
(58) (339)	separating digits (groups of three, from right to left)

the first digit is 3	$\sqrt[3]{58}$ is between 3 and 4
$\sqrt[3]{58{,}339}$ is between 30 and 40	add a zero for every digit triplet to the right of "58"
the number is between 35 and 40	58,339 is closer to 40^3 than to 30^3
39	$\sqrt[3]{58{,}339}$ (to the nearest unit)

Quiz Time

Here are 10 problems to test your skill in applying the concepts in this chapter. After attempting all 10 problems, read the explanations that follow. Then go back to the chapter and review your trouble spots. If you can handle the easier problems *and* the more challenging ones, consider yourself a "smart test-taker"!

Easier

1. Determine the precise values of (a), (b), and (c). Approximate the values of (d), (e), and (f) to the nearest integer.

(a) $\sqrt{225}$ (b) $\sqrt[3]{343}$ (c) $\sqrt{.0169}$ (d) $\sqrt{40}$ (e) $\sqrt{160}$ (f) $\sqrt[3]{-90}$

2. Combine terms in each of the following expressions. Express your answer in simplest form.

(a) $\sqrt{75} + \sqrt{12}$

(b) $\sqrt{125} - \sqrt{45}$

(c) $4\sqrt{27} - 2\sqrt{48} + \sqrt{147}$

3. Combine terms in each of the following expressions. Express your answer in the simplest form.

  (a) $\sqrt{9x}\,\sqrt{4x}$

  (b) $\sqrt{4}\,\sqrt{5}\,\sqrt{\frac{1}{2}}$

  (c) $\frac{1}{2}\sqrt{2}(\sqrt{6}+\frac{1}{2}\sqrt{2})$

4. Divide, expressing your answer in simplest terms.

  (a) $8\sqrt{12}\div 2\sqrt{3}$

  (b) $\sqrt{32b^3}\div\sqrt{8b}$

  (c) $42\sqrt{40x^3y^6}\div 3\sqrt{5xy^2}$

5. Simplify the following expression: $\sqrt[3]{\dfrac{81a^6b^3}{16c^5}}$

More Challenging

6. What is the value of $\sqrt{664}+\sqrt{414}$, to the nearest integer?

7. Simplify the following expression: $\sqrt{\dfrac{a^2}{b^2}+\dfrac{a^2}{b^2}}$

8. Simplify the following expression: $\sqrt{\dfrac{x^2}{36}+\dfrac{x^2}{25}}$

9. If $x < -1$, rank the following from largest to smallest in value:

  (a) $-\dfrac{.9}{x^3}$ (b) $-\sqrt[3]{x}$ (c) $\left(\sqrt[3]{x}\right)^2$ (d) $\dfrac{1}{\sqrt[3]{x}}$ (e) x^3

10. If $0 < x < 1$, rank the following from largest to smallest in value:

(a) $\sqrt{x}$ (b) $\sqrt{\dfrac{1}{x}}$ (c) $\sqrt[3]{x^2}$ (d) x^4 (e) $\dfrac{1}{x^2}$

Answers and Explanations

1. Here are the answer to the six problems:

(a)   15

(b)   7

(c)   .13

(d)   6 (closer to 6 than to 5)

(e)   13 (closer to 13 than to 12)

(f)   −4 (closer to −4 than to −5)

2. Here is the answer and analysis for each problem:

(a)   The answer is $7\sqrt{3}$

$$\sqrt{75} = \sqrt{25 \cdot 3} = 5\sqrt{3}$$
$$\sqrt{12} = \sqrt{4 \cdot 3} = 2\sqrt{3}$$
$$5\sqrt{3} + 2\sqrt{3} = 7\sqrt{3}$$

(b)   The answer is $2\sqrt{5}$

$$\sqrt{125} = \sqrt{25 \cdot 5} = 5\sqrt{5}$$
$$\sqrt{45} = \sqrt{9 \cdot 5} = 3\sqrt{5}$$
$$5\sqrt{5} - 3\sqrt{5} = 2\sqrt{5}$$

(c)   The answer is $11\sqrt{3}$.

$$4\sqrt{27} = 4\sqrt{9\cdot3} = 12\sqrt{3}$$
$$2\sqrt{48} = 2\sqrt{16\cdot3} = 8\sqrt{3}$$
$$\sqrt{147} = \sqrt{49\cdot3} = 7\sqrt{3}$$
$$12\sqrt{3} - 8\sqrt{3} + 7\sqrt{3} = 11\sqrt{3}$$

3.   Here is the answer and analysis for each problem:

(a)   The answer is $6x$.

$$\sqrt{9x}\sqrt{4x} = 3\sqrt{x}\cdot2\sqrt{x} = 6\left(\sqrt{x}\right)^2 \text{ or } 6x$$

(b)   The answer is $\sqrt{10}$.

$$\sqrt{4}\sqrt{5}\sqrt{\tfrac{1}{2}} = \sqrt{4\cdot5\cdot\tfrac{1}{2}} = \sqrt{10}$$

(c)   The answer is $\sqrt{3} + \tfrac{1}{2}$.

Distribute $\tfrac{1}{2}\sqrt{2}$ to each of the other two terms:

$$\tfrac{1}{2}\sqrt{2\cdot6} + \tfrac{1}{4}\sqrt{2\cdot2} = \tfrac{1}{2}\sqrt{4\cdot3} + \tfrac{1}{4}\sqrt{4} = \sqrt{3} + \tfrac{1}{2}$$

4.   Here is the answer and analysis for each problem:

(a)   The answer is 8.

$$\frac{8\sqrt{12}}{2\sqrt{3}} = 4\sqrt{4} = 4\cdot2 = 8$$

(b)   The answer is $2b$.

$$\frac{\sqrt{32b^3}}{\sqrt{8b}} = \sqrt{\frac{32b^3}{8b}} = \sqrt{4b^2} = 2b$$

(c)   The answer is $28xy^2\sqrt{2}$

$$\frac{42\sqrt{40x^3y^6}}{3\sqrt{5xy^2}} = \frac{42}{3}\sqrt{\frac{40x^3y^6}{5xy^2}} = 14\sqrt{8x^2y^4} = 28xy^2\sqrt{2}$$

5. The answer is $\dfrac{3a^2b}{2c}\sqrt{\dfrac{3}{2c^2}}$.

Separate out factors that are "cubes" of "easy" numbers. Then remove those factors and cubed variables from the radical:

$$\sqrt[3]{\frac{81a^6b^3}{16c^5}} = \sqrt[3]{\frac{(27)(3)a^6b^3}{(8)(2)c^5}} = \frac{3a^2b}{2c}\sqrt{\frac{3}{2c^2}}$$

6. The answer is 46. $\sqrt{664}$ ٩ $\sqrt{41f}$
                                          $\overline{25}$

There is no need to calculate either root precisely since the question asks for an approximation. $\sqrt{625}$ is 25, while $26^2 = 676$. 664 is closer in value to 676 than to 625. Thus, $\sqrt{664}$ is closer to 26 than to 25. $\sqrt{400} = 20$, while $21^2 = 44$. Since 414 is closer to 400 than to 441, $\sqrt{414}$ is closer to 20 than to 21. Thus, the sum of the terms is approximately 46 (20 + 26). (For the record, $\sqrt{664} \approx 25.8$, and $\sqrt{414} \approx 20.3$. Thus, their sum is approximately 46.1)

7. The answer is $\dfrac{a}{b}\sqrt{2}$.

You must first combine the two terms inside the radical, using the common denominator b^2 :

$$\sqrt{\frac{a^2}{b^2} + \frac{a^2}{b^2}} = \sqrt{\frac{a^2 + a^2}{b^2}} = \sqrt{\frac{2a^2}{b^2}} = \frac{a}{b}\sqrt{2}$$

8. The answer is $\dfrac{x}{30}\sqrt{61}$. You cannot move either term out of the radical without first combining them, using a common denominator:

$$\sqrt{\frac{x^2}{36} + \frac{x^2}{25}} = \sqrt{\frac{25x^2 + 36x^2}{36 \cdot 25}} = \sqrt{\frac{61x^2}{36 \cdot 25}} = \frac{x}{30}\sqrt{61}$$

$\begin{aligned}3&6\\2&5\\6&1\end{aligned}$

9. From largest to smallest (left to right on the number line), the order is: (b), (a), (d), (c), (e).

(b) must be a positive number greater than 1

(a) must be a positive non-integer between zero and 1.

(d) must be a negative non-integer between zero and -1

(c) must be greater than (to the right of) x, between x and -1 on the number line

(e) must be smaller than x (to the left of x on the number line)

If you wish to confirm this analyze, let $x = -2$, and using this value in each of the five expressions.

10. From largest to smallest, the order is: (e), (b), (a), (c), (d).

Let $x = \frac{1}{2}$. Using this value in each of the five expressions:

(a) $\sqrt{\frac{1}{2}} \approx \frac{1}{1.4}$ or $.71$

(b) $\sqrt{\frac{1}{\frac{1}{2}}} = \sqrt{2} \approx 1.4$

(c) $\sqrt[3]{\left(\frac{1}{2}\right)^2} = \sqrt[3]{\frac{1}{4}} \approx \frac{1}{1.6}$ or $.625$

(d) $\left(\frac{1}{2}\right)^4 = \frac{1}{16}$

(e) $\dfrac{1}{\left(\frac{1}{2}\right)^2} = \dfrac{1}{\frac{1}{4}} = 4$

8

Test-Taking Tips
for Improving Speed and Accuracy

So you think you're a pretty smart test-taker, having mastered Chapters 3 through 7? Well, it's one thing to understand math. It's quite another thing, however, to solve math problems quickly and accurately under time pressure during an important exam. Many people who are "good at math" are nevertheless not good at taking standardized math tests. So-called "book" knowledge does not make one immune to careless errors or a too-slow pace, either of which can prevent you from gaining admission to the school of your choice. So here in Chapter 8, as well as in Chapter 9, we're going to be taking a break from the "science" of math—all those boring rules and formulas—to explore the "art" of taking math tests. In this chapter, **you'll learn** how to:

- approximate and estimate numbers to help you solve problems quickly

- check your calculations to ensure against careless errors

- use a calculator to your advantage (if you're taking the SAT I or ACT)

We'll conclude the chapter with a list of 8 tips for taking standardized math tests. Because this chapter doesn't involve specific types of math problems, you won't be taking a quiz at the end of the chapter. In Chapter 9, you'll explore arithmetic shortcuts and tricks to help make calculations quicker and easier.

Approximating Numbers (Rounding Off)

The ability to estimate and approximate numbers is one of the most useful skills for taking standardized math tests; ironically, it is probably also the most neglected. Also ironic is that mathematics, by its very nature, involves precision and exactitude; yet in multiple-choice math questions, often you can identify the correct answer choice without determining the precise solution to the problem posed. Let's see how estimating and approximating numbers can make you a smarter test-taker.

How to Round off a Number

When you round off a number, you eliminate one of more of the digits from the right end of the number. This doesn't mean that you ignore those digits, however. Consider this not-so-round number:

$$4{,}834.826$$

If we simply ignore all digits to the right of the decimal point, we're left with 4,834. But we have not rounded off the number to the nearest unit, because .876 is closer to 1 than to 0. In this case, rounding off this number to the nearest unit requires increasing the "ones" digit from 4 to 5:

$$4{,}835$$

Just for the record, here's the same number rounded off to all possible places:

4,834.826 rounded to:	equals:	what we've done:
the nearest hundredth	4,834.83	round .826 up to .830
the nearest tenth	4,834.8	round .82 down to .80
the nearest unit	4,835	round 4.8 up to 5.0
the nearest ten	4,830	round 34 down to 30
the nearest hundred	4,800	round 834 down to 800
the nearest thousand	5,000	round 4,834 up to 5,000
the nearest ten thousand	0	round 4,834 down to 0

Did you notice the absence of the number 5 among the digits in the preceding number? Well, "5" requires explanation. It's midway between 0 and 10. So do you round it up or down? The answer is: it doesn't matter. The conventional method is to round all fives *up*. (I suppose mathematicians are optimists, preferring to view the proverbial glass as half full.) Here are some numbers with fives that are rounded up:

6,545.5 rounded to the nearest *unit* is 6,546

6,545 rounded to the nearest *ten* is 6,550

6,450 rounded to the nearest *hundred* is 6,500

6,500 rounded to the nearest *thousand* is 7,000

Don't worry: the correct response on your standardized exam will not depend solely on whether you round fives up or down. For example, if the solution to a problem is 4.5, you won't be asked to choose between 4 and 5 as the closest approximation.

Numbers with Never-Ending Decimal Places

Many numbers, especially square roots, include an infinite number of non-repeating decimal places. Such numbers are as "unround" as numbers can possibly be. $\sqrt{2}$, $\sqrt{3}$, and π are three good examples—they appear with particular frequently on math exams because they are essential to certain basic geometry formulas. So how should you handle them? On the exams, such numbers are usually expressed "as is" rather than as their decimal or fractional equivalents in the answer choices, so you won't have to deal with their values at all. Here's a simple multiple-choice illustration:

Q. If $12y^2 = x^2$, then $x =$

(A) $3y\sqrt{2}$ (B) $6y$ (C) $2\sqrt{3y}$ (D) $2y\sqrt{3}$ (E) $3y + \sqrt{2}$

The correct answer is (D). $x = \sqrt{12y^2}$ or $\sqrt{(3)(4)y^2}$. $4y^2$ can easily be removed from under the radical and expressed as it's square root, $2y$. This leaves only the number 3 under the radical. Notice that we did not have to estimate the value of $\sqrt{3}$ to determine the correct answer. The choices are expressed *in terms of* numbers such as $\sqrt{2}$ and $\sqrt{3}$. In other words, these numbers are left "as is."

If an exam question *does* require you to estimate the values of such numbers to determine the correct response, rounding them to the *nearest tenth* will usually suffice in this simple problem:

Q. Which of the following most closely approximates the area of a circle whose radius is 3?

(A) 18   (B) 27   (C) 28   (D) 30   (E) 36

The correct response is (C). The area of a circle is equal to πr^2, where r represents the radius. To the nearest tenth, $\pi = 3.1$. Substituting 3.1 for π:

$$A = (3.1)(3^2)$$

$$A = (3.1)(9)$$

$$A = 27.9$$

Are you skeptical about rounding off like this? Okay, let's use a more precise value for π. To the nearest hundredth, $\pi = 3.14$. Using this value:

$$A = (3.14)3^2$$

$$A = (3.14)(9)$$

$$A = 28.26$$

Again, the closest value among the answer choices is (C). "Wait a minute!" you say. What if the two closest values among the answer choices were 28 and 28.5? Using 3.1 as an approximate value of π would have resulted in the wrong answer:

27.9 is closer to 28 than to 28.5

28.26 is closer to 28.5 than to 28

Well, you probably won't be required to slice the numbers this fine on your exam. The test makers do not design questions to trick you in this manner. In the unlikely event that you do see answer choices this close in value, use a more precise approximation of the appropriate number (whether it be π, $\sqrt{2}$, or something else).

Remember: Always check the answer choices for clues as to how closely you must approximate numbers in solving the problem.

What about using fractional values to approximating never-ending numbers? For example, the fraction $\frac{22}{7}$ is often used as an approximation of π. Using this value would indeed have worked in the previous problem:

$$A = \left(\frac{22}{7}\right)(3^2)$$

$$A = \left(\frac{22}{7}\right)(9)$$

$$A = \frac{198}{7} \text{ or } 28\frac{2}{7}$$

In this case, however, you'll probably agree that the arithmetic is easier using a decimal approximation than a fractional one.

Remember: Use fractional approximations whenever the arithmetic is easier or when the answer choices are expressed as fractions. Otherwise, use decimal approximations.

Rounding off Numerators and Denominators

Nowhere does rounding off come more valuable than in dealing with fractions. Assume, for example, that an exam question involves the following division:

$$47 \div 62$$

You can't factor out any common divisors here, since 47 is a prime number. So are you stuck with performing long division? If so, it's a long way to the precise quotient:

$$62\overline{)47} = .758064$$

Don't worry; this is not the sort of number that you'll be required to compute on a standardized math exam. Smart test-takers realize that the test *maker* is not trying to gauge your ability to perform long division. In such cases, an approximations will almost certainly suffice. Whenever you're about to engage in this sort of arithmetical shenanigans, stop yourself in you tracks and think "rounding off" instead. Ask yourself these two questions:

Should I round these numbers up or should I round them down?

How far up or down should I round them?

Okay, you could round 47 down to 45 (multiples of 5 are often easy to work with) or up to 50 (a real nice round number); similarly, you could round 62 up to 65 or down to 60. Let's try each combination to see what quotients we come up with. In all cases, we'll express the problem as a fraction and "factor out" a 5 from both the numerator and the denominator:

1. $\dfrac{45}{65} = \dfrac{9}{13}$ 2. $\dfrac{45}{60} = \dfrac{3}{4}$ (or .75)

3. $\dfrac{50}{65} = \dfrac{10}{13}$ 4. $\dfrac{50}{60} = \dfrac{5}{6}$

Notice that our estimate using either #2 or #3 is closer to the precise value than our estimate using either #1 or #4. There's a good reason for this. In #2 and #3, we rounded both numbers *in the same direction* (either up or down), minimizing the change to the value of the overall fraction. #1 and #4 are not good approximations. Both distort the size of the fraction: #1 shrinks the fraction, while #4 enlarges it.

What if the answer choices are expressed in terms of decimals rather than fractions? $\dfrac{3}{4}$ is easily enough converted to .75, but what is $\dfrac{47}{62}$ in decimal terms? Oh, no, another division problem! You might want to use rougher approximations in such a case to avoid fractions. Specifically:

47 is just one unit less than 48

62 is just 2 units more than 60

$\dfrac{48}{6} = 8$ (no fractional values to mess with)

$\dfrac{48}{60} = .8$

Okay, we've distorted our estimate somewhat by rounding the numerator in a direction different from that of the denominator; but not by very much, and we came up with an approximation (.8) that's easy to work with.

So which method you use to approximate fractions depends on how easy it is to divide the revised numerator by the denominator. To underscore this important idea, let's look at the fraction $\dfrac{39}{83.8}$:

$$\frac{39\uparrow}{83.8\uparrow} \approx \frac{40}{85} = \frac{8}{17}$$

$$\frac{39\downarrow}{83.8\downarrow} \approx \frac{35}{80} = \frac{7}{16}$$

$$\frac{39\uparrow}{83.8\downarrow} \approx \frac{40}{80} = \frac{1}{2}$$

$$\frac{39\downarrow}{83.8\uparrow} \approx \frac{35}{85} = \frac{7}{17}$$

The first two estimates are more accurate than the last two, but the third one is the easiest to work with and might suffice if the answer choices are expressed only to the nearest half unit.

Remember: Round off the numerator of a fraction in the same direction as the denominator, unless the answer choices allow for a rougher approximation.

Techniques For Checking Your Calculations

Even the best and brightest number crunchers make careless errors from time to time. The kinds of errors we're referring to here are not mistaken notions about mathematical concepts, but rather those "oops" errors we all make from time to time in manipulating numbers and other terms. These "oops" errors may be due to momentary lapses in concentration or to the pressure of a timed-exam environment. Whatever their cause, more math questions on standardized tests are answered incorrectly due to "oops" errors than to any other factor. Take this as your cue: check your work before you move on to the next question.

"Checking your work" does not mean going through all the steps in solving the problem a second time. This isn't a very effective way to check your work, for two reasons:

1. In performing the same steps in the same sequence, you're more likely to repeat the same mistake than if you use some other approach.

2. Unless you move along at a very fast pace, you won't have time to work every problem twice; and if you do move at an extremely quick pace, you'll probably make many, many errors along the way.

What "checking your work" should entail is using one or more of the shorter techniques discussed below to make sure you haven't inadvertently committed an "oopsie."

Reversing the process

The most obvious (but nevertheless commonly overlooked) method of checking arithmetical operations is to reverse the computational process:

If you've:	check your work by:
added two numbers together $56 + 233 = 289$	subtracting one of the numbers from the sum $289 - 233 = 56$ or $289 - 56 = 233$
subtracted one number from another $28.34 - 3.8 = 24.54$	adding the result to the number you subtracted $24.54 + 3.8 = 28.34$
multiplied two numbers $1.13 \times 6.65 = 81.396$	dividing the product by one of the numbers $81.396 \div 1.13 = 6.6$ or $81.396 \div 6.65 = 1.13$
divided one number by another $789 \div 3 = 263$	multiplying the quotient by the second number $263 \times 3 = 789$

The "Sum of the Digits" Approach to Checking Your Arithmetic

Here's a strange but wonderful technique that can be used to check your work for most arithmetic problems. It's probably most useful for checking multiplication. The technique involves adding up the digits of each number you're working with.

Addition and Subtraction

Add all the digits of your sum, then add the digits of this sum, and keep going until you're left with a single-digit number. Do the same for each of the numbers you added together in the original problem. The first digit you determined by this method should equal the sum of the other digits. If not, you made a mistake in your addition. Here's how it works:

original problem:	$8{,}548 + 2{,}935 = 11{,}438$
add the digits of the sum:	$1 + 1 + 4 + 8 + 3 = 17$
add the digits of this sum:	$1 + 7 = 8$
add the digits of 8,548:	$8 + 5 + 4 + 8 = 25$
add the digits of this sum:	$2 + 5 = 7$
add the digits of 2,935:	$2 + 9 + 3 + 5 = 19$
add the digits of this sum:	$1 + 9 = 10$
add the digits of this sum:	$1 + 0 = 1$
checking the final sums:	$7 + 1 = 8$

You can speed up this process further by ignoring (or "casting out") all the 9's as you add digits together—for the number 2,935 in the example above:

$$2 + 3 + 5 = 10 \quad \text{(ignoring the digit "9")}$$
$$1 + 0 = 0$$

You can also use the "sum of the digits" approach to check subtraction, but it's probably more trouble than its worth, since it's easy to add numbers together to check subtraction.

Multiplication

You can also us the "sum of the digits" approach to check multiplication. It works exactly the same as with addition, except that in the final step you *multiply* your single digits. If the numbers you're working with include decimals, just ignore them, as illustrated here:

original problem	$23 \times 7.26 = 166.98$
add the digits of 166.98 (ignore the 9 as well as the 1 and 8, since $1 + 8 = 9$)	$6 + 6 = 12$
add the digits of this new sum	$1 + 2 = 3$
add the digits of 23	$2 + 3 = 5$
add the digits of 7.26	$7 + 2 + 6 = 15$
add the digits of this new sum	$1 + 5 = 6$
checking the final digits with multiplication (ignore any incorrect decimal places)	$5 \times 6 = 3 \quad (5 \times 6 = 30)$

Division by 9

The "sum of the digits" approach can be used to check division by 9 (it doesn't work for other numbers). Divide a number by 9. Then, compare your remainder with the sum of the digits of the number you divided by. (As with addition, repeat the process until you have a single-digit number.) Here's an example:

$$3{,}677 \div 9 = 48\frac{5}{9} \quad \text{(the remainder is 5)}$$

$$3 + 6 + 7 + 7 = 23$$

$$2 + 3 = 5$$

$$5 = 5$$

As with addition, in adding up the digits you can ignore all 9's; for example:

$$698 \div 9 = 77\frac{5}{9}$$

The remainder is 5

$$6 + 8 = 14 \text{ (ignoring the digit "9")}$$

$$1 + 4 = 5$$

$$5 = 5$$

Keep in mind, however, that using the sum-of-the-digits approach for division by 9 does not serve to check the accuracy of your quotient (77 in this case).

A Few Other Ways to Check Your Arithmetic

If you just don't have time to reverse your computation as suggested above, you may still have time to use one of these two shortcut methods for checking arithmetic:

- Recalculate the smallest digit (the one furthest to the right) to make sure it's correct (it's quick and easy, since you don't have to think about carried numbers)

- If you multiplied two numbers, check your work with cross-multiplication. (You'll learn about this shortcut multiplication technique in Chapter 9).

Using Calculators To Your Advantage (On the SAT I or ACT)

The use of calculators is permitted on SAT I and ACT, but not on other standardized math exams (such as the GMAT and the GRE). So if you're preparing for SAT or ACT, read this section; otherwise, you can skip this section and proceed to Chapter 9.

You may bring to the exam session any of the following types of calculators:

- four-function
- scientific
- graphing

You may not bring calculators of the following types:

- calculators with paper tape or printers
- laptop computers
- pocket organizers
- hand-held microcomputers

Make sure you are thoroughly familiar with the calculator you bring to the exam.

No SAT or ACT question requires the use of a calculator. For some questions a calculator may be helpful. For other questions, however, a calculator may be inappropriate or may actually slow you down. In any event, remember that a calculator is only a tool to avoid inaccuracies in computation; it cannot take the place of understanding how to set up and solve a mathematical problem.

Here's a sample problem for which a calculator *might* be helpful:

> Q. **The price of one dozen roses is $10.80. At this rate, what is the price of 53 roses?**

The simplest way to approach this question is to divide $10.80 by 12, which gives you the price of one rose, then multiply that price by 53:

$$\$10.80 \div 12 = \$.90$$

$$\$.90 \cdot 53 = \$47.70$$

Although the computations are fairly simple, using a calculator might improve your speed and accuracy. Let's change the problem so that a calculator would not be useful:

> Q. **If the price of x roses is D dollars, what is the price of y roses?**
>
> (A) Dx (B) $\dfrac{xy}{D}$ (C) $\dfrac{y}{Dx}$ (D) $\dfrac{Dy}{x}$ (E) $\dfrac{Dx}{y}$

In this version of the problem, use the proportion method to set up an equation, letting S equal the solution. Then solve for S:

$$\frac{x}{D} = \frac{y}{S}$$

$$Sx = Dy$$

$$S = \frac{Dy}{x}$$

This is really a pure algebra problem, requiring no numerical calculations. A calculator would be useless in solving problems such as this one.

In some questions involving numbers, although it might be possible to use a calculator to help solve the problem, doing so might slow you down because the question is set up to be solved in a quicker, more intuitive manner. Here's an example:

> Q. If $x = \dfrac{1}{2} \cdot \dfrac{1}{3} \cdot \dfrac{3}{2} \cdot \dfrac{4}{81}$, then $\sqrt{x} =$

In this problem, you *could* use a calculator to perform all the steps:

1. multiply the numerators

2. multiply the denominators

3. divide the product of the numerators by the product of the denominators

4. compute the square root

However, using a calculator is far more trouble than it's worth here. The problem is set up so that all numbers but 81 cancel out:

$$x = \frac{1 \cdot 1 \cdot 3 \cdot 4}{2 \cdot 3 \cdot 2 \cdot 81} = \frac{1}{81}$$

The problem is easily (and more quickly) solved this way, without a calculator:

$$\sqrt{\frac{1}{81}} = \frac{1}{9}$$

8 Tips For Smart (Math) Test-Takers

1. Size up the question first.

Try to assess what specific mathematical area is being covered (for example, what mathematical rules and formulas come into play). Every question is included in the exam because it tests you on one or two specific rules or formulas. By determining up front what you're up against, you're already well on your way to dealing with the question.

2. Determine how much time, if any, you're willing to spend on the problem.

Math problems on standardized exams vary widely in difficulty level, from simple arithmetic calculations to complex word problems. As you work through the remaining chapters of this book, try to get a feel for your limitations in handling complex questions. Learn to recognize a "toughie" question when you see it, so that you don't waste valuable time on it; take an educated guess, and move on.

3. Size up the answer choices.

Before you attempt to solve a particular multiple-choice question, examine the answer choices. The answer choices often provide helpful clues about how to proceed in solving the problem and about what sort of solution you should be looking for as you work. Pay particular attention to the following:

Form. Are the answer choices expressed as percentages, fractions, or decimals? Ounces or pounds? Minutes or hours? If the answer choices are expressed as equations, are all variables together on one side of the equation? As you work through the problem, convert numbers and expressions to the same form as the answer choices.

Size. Are the answer choices extremely small numbers? Numbers between 1 and 10? Larger numbers? Negative or positive numbers?

Variation in Size. Do the answer choices vary widely in value, or are their values clustered closely around an average? If all answer choices are tightly clustered in value, you can probably disregard decimal points and extraneous zeros in performing calculations. At the same time, however, you should be more careful about rounding off your figures where answer choices do not vary widely. Wide variation in value suggests that you can easily eliminate answer choices that don't correspond to the general size of number suggested by the question.

Other distinctive properties and characteristics. Are the answer choices integers? Do they all include a variable? Do one or more include radicals (roots)? Exponents? Is there a particular term, expression or number that they have in common?

4. Don't do too much work in your head.

Carelessness, *not* lack of knowledge or ability, is the leading cause of incorrect responses on standardized math exams. When you perform calculations and manipulate expressions, use your pencil and scratch paper for all but the simplest mathematical steps.

5. On multiple-choice questions, beware of "sucker bait" choices.

The test makers intentionally "bait" you with wrong-answer choices that result from making specific common errors in setting up problems and in calculation. Don't be a sucker by assuming that your response is correct just because your solution appears among the five answer choices! Rely instead on your sense for whether you understood what the question calls for and performed the calculations and other steps carefully and accurately.

6. Don't do more work than is necessary to determine the correct answer.

If the question asks for an approximation, the test maker is providing a hint that exact calculations may be time consuming and that you can rely on rounded-off numbers as you work. Even if the question does not explicitly call for approximation (and the overwhelming majority of questions do not), don't calculate exact figures when approximations will suffice.

7. Verify that your answer is in the right ballpark.

We all fail at times to see the forest for the trees, becoming so engrossed with details that we lose sight of the "big picture." It's easy to fall into this trap with math problems. A number cruncher can easily lose sight of what he or she is attempting to compute. Although rounding of decimal places and checking your arithmetic is useful, these methods do NOT ensure that your solution is in the right "ballpark" in terms of its *size* and *form*. After solving the problem and selecting a response, check the question again to verify that your response corresponds to what the question calls for—in terms of size, expression, units of measure, and so forth. If it does, and you are confident that your work was careful and accurate, don't spend any more time checking your work. Confirm your response and move on to the next question.

8. Stay on pace, but don't be a clock-watcher.

Try to move at a steady pace without feeling rushed, even if it means not finishing all of the questions in the allotted time. It is distracting and time-consuming to keep track of your elapsed time for each question. You should, however, check the clock every so often—perhaps every 10 minutes—to see whether you are "on pace."

7. Verify that your answer is in the right ballpark.

We all fail at times to see the forest for the trees, becoming so engrossed with details that we lose sight of the "big picture." It's easy to fall into this trap with math problems. A number cruncher can easily lose sight of what he or she is attempting to compute. Although rounding of decimal places and checking your arithmetic is useful, these methods do NOT ensure that your solution is in the right "ballpark" in terms of its *size* and *form*. After solving the problem and selecting a response, check the question again to verify that your response corresponds to what the question calls for—in terms of size, expression, units of measure, and so forth. If it does, and you are confident that your work was careful and accurate, don't spend any more time checking your work. Confirm your response and move on to the next question.

8. Stay on pace, but don't be a clock-watcher.

Try to move at a steady pace without feeling rushed, even if it means not finishing all of the questions in the allotted time. It is distracting and time-consuming to keep track of your elapsed time for each question. You should, however, check the clock every so often—perhaps every 10 minutes—to see whether you are "on pace."

9

Not-So-Stupid
Arithmetic Tricks

ave you ever wondered how certain people, such as those math whizzes who grace television info-mercials and talk shows with their amazing numeric prowess, can compute numbers quickly in their heads, seemingly without effort? Are they natural geniuses? Probably not. Have they memorized their "times" tables to the thousands place value? Certainly not. Aside from the rare savant (remember the movie *Rainman*?), what so-called "math whizzes" have done is to learn and practice a handful of techniques that enable them to perform computations quicker than you can using conventional methods.

You'll learn many of these shortcuts in this chapter. Some of these techniques will help you get to the correct answers on timed exams more quickly. Others are less pertinent to standardized exams but will be quite useful in other situations. Keep in mind that these techniques depart from the conventional ways of performing arithmetic. As a result, **you need to practice** them extensively before you're ready to put them to use on a test.

Expressing 25, 50, and 75 in Terms of 100

This is a simple but extremely useful technique for speeding up multiplication involving the numbers 25, 50, and 75. When crunching these numbers, think about them in this way:

$$25 \text{ is } \frac{1}{4} \text{ of } 100$$

$$50 \text{ is } \frac{1}{2} \text{ of } 100$$

$$75 \text{ is } \frac{1}{2} \text{ of } 100 + \frac{1}{4} \text{ of } 100$$

Let's look at one example of each.

The number 25

Quick...perform the following operation in your head:

$$\begin{array}{r} 25 \\ \times 32 \\ \hline \end{array}$$

How long did it take you to determine that the answer is 800? If you solved the problem the conventional way, you did far too much work. Try one of these two shortcut methods instead:

1. 25 is $\frac{1}{4}$ of 100. So...multiply 100 by 32 (that's easy), then divide the product (3,200) by 4 (that's easy). Both calculations are no-brainers, and you're done in a millisecond!

2. Divide 32 by 4 (the same thing as multiplying 32 by .25), then multiply the result, which is 8, by 100. Voila!

Why were these alternative methods easier? Well, in most cases, it's easier to divide a number by 4 than to multiply it by 25.

Note: You can also use this shortcut with the number 12.5, converting it to 100 by multiplying by 8. (12.5 is $\frac{1}{2}$ of 25.)

The number 50

You can apply the same technique, of course, to the number 50:

$$\begin{array}{r} 50 \\ \times 82 \\ \hline \end{array}$$

50 is $\frac{1}{2}$ of 100. So...instead of multiplying 5 by 82 and tacking on a zero, simply divide 82 by 2 (41) and add the same zero (410).

Remember: It's usually quicker and easier to divide big numbers by 2 than to multiply them by 5.

The number 75

Dealing with the number 75 is a two-step process. Quick...compute this:

$$\begin{array}{r} 75 \\ \times 16 \\ \hline \end{array}$$

The shortcut here is to think of 75 in one of these two ways:

$$100 - \frac{1}{4} \text{ of } 100$$

$$50 + \frac{1}{2} \text{ of } 50$$

75×16

With the first method, start with 100 instead of 75. $100 \times 16 = 1,600$. Now, subtract 400 ($\frac{1}{4}$ of 1,600). The answer is 1,200. With the second method, start with 50 instead of 75. $\frac{1}{2}$ of 16 is 8; adding back two zeros, $50 \times 16 = 800$. Add $\frac{1}{2}$ of 800 to 800. Again, the total is 1,200.

Mental Multiplication with the "Round-Off Two Step"

No, we're not talking about a new square dance here. We're talking about performing multiplication as efficiently as possible. It's much easier to work with numbers like 100, 50, or 10 than with numbers like 98, 53, or 11, isn't it? In the real world (including real standardized exams), however, all numbers don't end in zero. But you can always pretend...and get away with it. Here's how. Quick...perform the following operation in your head:

$$\begin{array}{r} 99 \\ \times 98 \\ \hline \end{array}$$

How long did it take? How many steps? If it took you more than 2 seconds, you're doing too much work! The key is to realize that multiplying any number by 99 is the same as adding it 99 times and that 99 times is *one time less than 100*. So...why not round 99 up to 100, then compensate for rounding up by subtracting the extra 98 from the product? Here are the two steps:

1. $100 \times 98 = 9,800$ (a no-brainer calculation)

2. $9,800 - 98 = 9,702$ (a semi-no-brainer calculation)

Let's see how you might take this technique a step further. Try crunching these numbers quickly in your head:

$$\begin{array}{r} 109 \\ \times 49 \\ \hline \end{array}$$

In multiplying these two numbers, did it occur to you that you can use rounding-off to your advantage *twice*? Here are the steps:

1. round 109 down to 100

2. add 9×49 to that total; first round 9 up to 10

3. subtract 49 from 490 to compensate for rounding up

4. add 441 to 4,900

1. $100 \times 49 = 4{,}900$

2. $10 \times 49 = 490$

3. $490 - 49 = 441$

4. 5,341 (your answer)

Warning: Don't try to round off more than one number at the same time here (100×50 in this example). Compensating for simultaneous rounding is more trouble than it's worth.

Remember: When facing a difficult number-crunching problem, always check to see if any of the numbers are close to "easy" numbers—especially numbers ending in zero. If so, round up or down, then compensate for it an a second step.

Multiplying Numbers That Are Close in Value

Rounding one number up and another one down by the same amount works quite nicely to speed up addition—you're just taking the average of the two numbers (more on this later). Does it work for multiplication? Well, yes and no. Here's what you *can't* get away with:

99×101

100×100 (increasing 99 by 1 and decreasing 101 by 1)

10,000

Wrong! $99 \times 101 = 9,999$, *not* 10,000. Oh...9.999 is an interesting number. It's just 1 less than 10,000. Hmm, is there a potential shortcut lurking here? Yes, there is. The table below reveals the pattern:

square of the two numbers averaged	difference of 1	difference of 4	difference of 9
$100^2 = 10,000$	$99 \times 101 = 9,999$	$98 \times 102 = 9,996$	$97 \times 103 = 9,991$
$60^2 = 3,600$	$59 \times 61 = 3,599$	$58 \times 62 = 3,596$	$57 \times 63 = 3,591$
$50^2 = 2,500$	$49 \times 51 = 2,499$	$48 \times 52 = 2,496$	$47 \times 53 = 2,491$
$25^2 = 625$	$24 \times 26 = 624$	$23 \times 27 = 621$	$22 \times 28 = 616$
$20^2 = 400$	$19 \times 21 = 399$	$18 \times 22 = 396$	$17 \times 23 = 391$
(you should memorize these "squares")	conclusion: subtract 1 from the square	conclusion: subtract 4 from the square	conclusion: subtract 9 from the square

Here's the shortcut that emerges: If two numbers differ by an even amount, instead of multiplying the two numbers, square the number in between them (the average of the two numbers). Then subtract the *appropriate number* from the square. What is the "appropriate number?" Well, it's 1 if the numbers differed by 2, it's 4 if the numbers differed by 4, and its 9 if the numbers differed by 6.

Go further if you'd like, with numbers that differ by 8, 10, 12, and so forth. I think you'll find that this technique is more trouble than it's worth, though, when the two numbers differ by more than 8.

This technique works with all numbers. However, when the two numbers differ by an odd amount (not an even integer), the computation is far more trouble than its worth. Also, when the number midway between the two original numbers is not a nice round number, it may be more time-consuming to square that number than to multiply the original two numbers. Applying our newly discovered "shortcut" to 65 and 69:

$$65 \times 69 = 67^2 - 4$$

So what is 67×67? It's probably just as easy to compute 65×69, isn't it? If you're looking for a shortcut that works with numbers such as these, pay close attention to the next section.

Multiplying Numbers Using a Round-Number Reference Point

The shortcut discussed in the previous section is not all that useful in combining numbers such as these:

$$\begin{array}{r} 94 \\ \times 98 \\ \hline \end{array}$$ (average is 96—not a nice round number)

$$\begin{array}{r} 98 \\ \times 97 \\ \hline \end{array}$$ (average is $97\frac{1}{2}$ —definitely not a nice round number)

So is there some other way to get around long multiplication here? Notice that the two numbers in each pair above are pretty close to each other in value, and they are also close to the same nice round number—100. These features are the keys to the shortcut multiplication method you'll learn right now. Here are the steps, using the first expression (94×98):

1. Think "100" (the nice round number a bit greater than the other two). 94 is less than 100 by 6, and 98 is less than 100 by 2.

2. Subtract 6 from 98 or subtract 2 from 94 (you get the same number either way). Write down the result:

92

3. Multiply together those two "differences" from the first step: $6 \times 2 = 12$. Write down this product the right of the other number:

92 12

That's the answer: 9,212! Go ahead and check it, using regular multiplication or a calculator. See? Keep in mind that in the last step, had the product been a *single*-digit number, we would have had to account for this by adding a zero to the left of it. To illustrate this point, let's perform the three steps above on the operation 98×97:

98×97	two numbers approaching 100
2 and 3	difference between 100 and each number
95	$98 - 3$ or $97 - 2$
6	2×3 (a single-digit product)
9,506	final answer (extra digit required)

This technique also works for numbers that exceed a nice round number by a little bit, except that instead of subtracting in the third step, you add. Here's how it works with 105×112:

105×112	two numbers a bit over 100
5 and 12	difference between 100 and each number
117	$105 + 12$ or $112 + 5$
60	5×12
11,760	final answer

You can also use this technique with larger numbers, such as those around 1,000 or 10,000. With some modifications (which we won't get into here), the technique also works in multiplying number pairs near "semi-round" numbers, such as 50 or 500.

What To Do with All Those Zeros?

Zeros...seems you can't live with them, but you can't live without them, either. Tag some zeros onto an otherwise innocuous little number, and you've created an unwieldy beast. On the other hand, in arithmetic it's usually easier to face a string of zeros than a string of other numbers. In any event, here are a few time-saving techniques will help you tame the wild "zero," no matter what mathematical jungle you find yourself in.

The "Zero Sum Game" (Multiplying Numbers with Trailing Zeros)

Do you panic when faced with multiplication involving big numbers ending in lots of zeros (we'll call them "trailing zeros")? Well, calm down; you just need to keep in mind a few simple tips. Let's take a look at an example:

$$4,200$$
$$\times 6,000$$

You know the product of these two numbers is big, but how big? Here's how to make sure you get it right:

1. ignore all the trailing (consecutive) zeros at the end of the numbers

2. multiply whatever numbers are left (in this case, $42 \times 6 = 252$)

3. add however many zeros you ignored to the end of your product (in this case, we ignored 5 zeros, so the answer is 252 with 5 zeros tagged onto the end: 25,200,000)

By the way: Among the ancient civilizations, only the Babylonians used a numbering system that included the concept of zero as a placeholder in larger numbers (such as 409). Even the Babylonian system, though, could not accommodate trailing zeros. (That's why average SAT scores were so low in ancient Babylon.)

Canceling Trailing Zeros in Division

Now, let's consider the same two numbers as above, but we'll divide one by the other instead:

$$\frac{4,200}{6,000}$$

Here's how to make sure you get it right:

1. Ignore the zeros, and divide what's left ($\frac{42}{6} = 7$)

2. Cancel (cross out) every trailing zero in the numerator for which there is also a trailing zero in the denominator (in this case, all zeros cancel out except for one zero in the denominator.

3. For every extra zero in the denominator, move the decimal point in your quotient to the left one place (in this case, take our quotient 7 and move the decimal point one place to the left—the answer is .7)

4. For each extra zero in the numerator, move the decimal place to the right one place (this doesn't apply here, but see below)

Let's try it again, this time reversing the numerator and the denominator:

$$\frac{6,000}{4,200}$$

1. Ignore the zeros, and divide what's left ($\frac{6}{42} = \frac{1}{7}$ or about .14)

2. Cancel trailing zeros in the numerator and denominator

3. You're left with one extra zero in the numerator, so move the decimal place to the right one place. The answer is about 1.4. (To express the answer as a fraction, we would add the extra zero to the numerator: $\frac{10}{7}$.)

Remember: You can ignore trailing zeros...but only temporarily. You'll have to tag them back onto your final number; and you can't ignore zeros between non-zero numbers (like 308).

Really Long Numbers and Scientific Notation

To round out your number-crunching education, let's think really big for a few moments. Quick...what is this number?

8,370,000,000,000,000,000.0000000000000000000000009

Well, it's "eight quintillion, three-hundred-seventy quadrillion, and nine septillionths" (or, if you want to express it the short way: "eight-*point-three-seven* quintillion, and nine septillionths"). Our names for really big numbers are based on the Greek language, and they change by three-digit increments (separated by commas), starting with one million. Here they are, through 28 digits:

1,000,000	6 zeros (2 sets of 3)	million ("mil" means one)
1,000,000,000	9 zeros (3 sets of 3)	billion ("bi" means two)
1,000,000,000,000	12 zeros (4 sets of 3)	trillion ("tri" means three)
1,000,000,000,000,000	15 zeros (5 sets of 3)	quadrillion ("quad" means four)
1,000,000,000,000,000,000	18 zeros (6 sets of 3)	quintillion ("quint" means five)
1,000,000,000,000,000,000,000	21 zeros (7 sets of 3)	sextillion ("sex" means six)
1,000,000,000,000,000,000,000,000	24 zeros (8 sets of 3)	septillion ("sept" means seven)
1,000,000,000,000,000,000,000,000,000	27 zeros (9 sets of 3)	octillion ("oct" means eight)

Mathematicians and scientists avoid all these zeros by using scientific notation, which simply indicates how many places a decimal point should be moved (either to the left or right). Great idea, huh? To express any number in scientific notation, use a number with exactly one digit to the left of the decimal point (like 2.88 or 5.9), and multiply the number by 10 raised to a particular power (exponent). For example, here's 8.37 quintillion in scientific notation:

$$8.37 \times 10^{18}$$

Notice that the exponent (18) corresponds to the number of zeros following one quintillion in the table above; that's no coincidence. The exponent determines where the decimal point goes—move the decimal point to the right (in this case, from 8.37) by the exact number of places that the exponent calls for.

What about really *small* numbers? How are they expressed in scientific notation? Well, simply by using a negative exponent.

Remember: A negative sign in an exponent does not change the sign of the base number; it merely reduces its size. Here's nine septillionth in scientific notation:

$$9.0 \times 10^{-24}$$

The exponent indicates how many places to move the decimal point left from 9.0. Go ahead and count these places in that big number at the beginning of this section. 24, right?

For a different perspective, here are some more down-to-earth numbers in scientific notation (notice on the next page that $10^{0} = 1$):

$$210{,}000 = 2.1 \times 10^{5}$$

$$47{,}200 = 4.72 \times 10^{4}$$

$$3{,}922.8 = 3.9228 \times 10^{3}$$

$$999 = 9.99 \times 10^{2}$$

$$82.2 = 8.22 \times 10$$

$$4.8 = 4.8 \times 10^0$$

$$.907 = 9.07 \times 10^{-1}$$

$$.0443 = 4.43 \times 10^{-2}$$

$$.007 = 7.0 \times 10^{-3}$$

Of course, scientific notation is more trouble than it's worth for numbers like the ones in this table that are neither extremely large nor extremely small. Indeed, it's of little practical use unless you're an astro-physicist or you're computing the net worth of Bill Gates (the president of Microsoft Corporation and wealthiest person in the world).

Did you know? A 1 followed by 100 zeros is called a *google*. (No, "google" is not a Greek word; physicist George Gammow coined the term when describing big numbers to his young daughter). In scientific notation, 1 google $= 1.0 \times 10^{100}$.

Cross-Multiplication

Don't confuse this type of cross-multiplication with the method used to eliminate fractions on both sides of an equation. What we're dealing with here is pure number crunching. Quick...try to mentally compute the product of these two numbers, without using any of the techniques discussed so far:

$$\begin{array}{r} 83 \\ \times 47 \\ \hline \end{array}$$

It's tough without a pencil, isn't it? Well, here's a little-known technique to help make mental multiplication of any two multi-digit numbers a bit easier. Think of the problem above as a matrix of four digits; your job is to multiply single digits *vertically* as well as *diagonally* (that's the "cross" part)—*four* operations altogether in this example. Here's how it works:

1. Multiply together the biggest place values (in this case, 8 and 4), adding one zero for each additional digit in each number (two zeros altogether here):

$$3,200$$

This is a great start—we already know the minimum size of the answer, don't we?

2. Multiply either two of the diagonal numbers: $8 \times 7 = 56$. Add one zero for each additional place (one altogether in this case):

$$560$$

3. Multiply the other two diagonal numbers: $4 \times 3 = 12$. Add one zero for each additional place (one altogether in this case):

$$120$$

Combine the totals from steps 2 and 3, adding them to the product in step 1:

$$560 + 120 = 680$$

$$680 + 3,200 = 3,880$$

4. Multiply the "ones" digits: $3 \times 7 = 21$. Add this to our running total for the final answer:

$$3,901$$

By the way, cross-multiplication also works from right to left (try it, if you like). Nothing about this technique is magic. All we're doing here is performing the steps in a different order from the one to which you are accustomed. Nevertheless, it's a better way to multiply numbers in your head. Of course, the more numbers you add and the more digits the numbers include, the tougher it gets to store the running totals in your head. It takes a lot of practice to be able to perform mental multiplication of triple-digit (and larger) numbers.

Remember: Practice left-to-right addition and multiplication a lot before trying it on timed exams; it takes practice to break old habits and become quick at using new techniques.

Left-to-Right Multiplication

Most of us learned to multiply multiple-digit numbers from right to left, carrying digits and adding sub-products along the way. Like addition, however, moving from left to right is actually more intuitive, especially for people who are more verbal than mathematical, and it's easier to multiply numbers in your head this way. The first step produces your biggest number—the one that establishes the overall size of the answer. Then you fill in the places, moving to the right with each step. Here's how it works. Quick...compute these two products in your head:

$$
\begin{array}{r} 36 \\ \times\ 7 \\ \hline \end{array}
\qquad\qquad
\begin{array}{r} 83 \\ \times\ 9 \\ \hline \end{array}
$$

Do yourself a favor and start with the left-hand ("tens") digit:

$30 \times 7 = 210$
add 42 (6×7) to 210
252

$80 \times 9 = 720$
add 27 (3×9) to 720
747

Okay, let try left-to-right multiplication with larger numbers:

$$
\begin{array}{r} 247 \\ \times\ \ 63 \\ \hline \end{array}
$$

Here are the steps for handling this problem from left to right:

1. First apply the 60 to 247. Round up 247 to 250 and multiply by 60—that's 25×6 with two places to right; that's how many zeros you ignored (more about ignoring zeros later in this chapter):

$$15,000$$

Since 250 exceeds 247 by 3, compensate for rounding up by subtracting 180 (60×3) from 15,000 (we're using the "round-off two-step" from the previous section here):

$$14,820$$

2. Next, apply the 3 to 247, again, rounding 247 up to 250, then compensating in a second step (the "round-off two step again):

$$3 \times 250 = 750$$
$$750 - 9 = 741$$

3. Finally, add 741 to 14,820 (carrying a 1 increases 14*** to 15***):

15,561 (your answer)

Left-to-Right Addition

Quick...add together these three numbers in your head, *without* a pencil or calculator:

$$675$$
$$986$$
$$\underline{+733}$$

It's pretty tough to keep all the numbers in your head as you add up each column, isn't it? Instead of starting with the "ones" column and moving from right to left, effective number-crunchers move from left to right. Here's how to do it, using the problem above as an example:

1. Starting with the hundreds place value, add up the digits ($6 + 9 + 7 = 22$), leaving two places to the right of the total:

22**

2. Add up the "tens" digits ($7 + 8 + 3 = 18$), and increase the "hundreds" value from 22 to 23 accordingly:

238*

3. Add up the "ones" digits $(5 + 6 + 3 = 14)$, and increase the "tens" value from 8 to 9 accordingly:

$$2,394$$

What we've done here is not magic, by any means. It's just a better way to add numbers in your head. There are three reasons for this:

- You don't have to remember as many numbers along the way. In our example, we only had to remember two numbers—22 and 238—as we went, and we only had to remember one of these at a time.

- Left-to-right addition is more intuitive because we say numbers and think about them from left to right. For example, we don't think about 2,394 as 4 units, 90 more, 300 more, and 2,000 more. We think about it as 2,000, with 300 and 90 and 4 added to it.

- Left-to-right addition determines the overall size of the number right away. In our example, we know immediately after the first step that the total must be greater than 2,200 and less than 2,500. (In fact, the number can be 2,497 at the most. This maximum value assumes that all "tens" and "ones" digits are maximized at 9.)

So why did we all learn to add columns of numbers from left to right? Simple. Because we wrote down the column totals, and moving from right to left, you don't have to erase and revise your totals as you go, as you do moving from left to right. Here's the upshot: if you're using a pen or pencil, add numbers the old-fashioned way—from right to left. If you're adding numbers in your head, go from left to right.

Quiz Time

Here are 10 problems to test your skill in applying the concepts in this chapter. After attempting all 10 problems, go back to the chapter and review your trouble spots. If you can handle these problems with the techniques discussed in this chapter, consider yourself a "smart test-taker"!

1. Multiply these numbers by first converting the bottom number to 100 (or a multiple of 100). Check your answers with a calculator.

66	341	7791	120	720
×25	×50	× 25	×75	×125

2. Multiply these numbers using the two-step round-off technique. Check your answers with a calculator.

97	59	121	7	519
×99	×60	×12	×25	×19

3. Multiply these number pairs by squaring the number midway between them and subtracting the appropriate number. Check your answer with a calculator.

31	37	68	11	154
×29	×43	×72	×109	×146

4. Multiply these numbers using a round-number reference point. Check your answers with a calculator.

58	76	289	108	128
×73	×76	×294	×109	×126

5. Solve the following five problems, using the techniques for handling trailing zeros that you learned in this chapter. Check your answers with a calculator.

30	710	800	300	61,000
×90	×80	×140	×900	×5,900

6. Express the following five fractions as decimal numbers, using the techniques for handling trailing zeros that you learned in this chapter. Check your answers with a calculator.

$$\frac{30}{90} \qquad \frac{80}{710} \qquad \frac{800}{140} \qquad \frac{300}{900} \qquad \frac{61,000}{5,900}$$

7. Multiply these numbers using cross-multiplication. Check your answers with a calculator.

29	87	63	52	91
×74	×17	×45	×68	×37

8. Try using left-to-right multiplication to solve these five problems. Check your answers with a calculator.

24	76	67	45	219
×3	×7	×38	×86	×374

9. Try using left-to-right addition to solve these five problems. Check your answers with a calculator.

69	58	959	497	1,475
+45	+79	+80	+239	+8,747

10. Just for fun, test your skill at "old-fashioned" multiplication. What are the first two 3-digit numbers in the following problem, assuming that the number 7 does not appear among *any* of the digits in the problem? There's only one possibility.

```
  2XX
  3XX
  5XX
 X4X
 XX3
XXXXX
```

(Answer to #10: The two numbers are 281 and 332)

10

Algebra
—Basic Concepts

Does the mere mention of the word "algebra" send you screaming into the night? Do you suffer from recurring nightmares in which the whole world is laughing at you while you try in vain to simply "solve for x"? If so, the next three chapters were written for you. Chapter 10 covers the basic skills you'll need to handle any algebra problem on a math exam.

First, **we'll review the rules** for combining and simplifying algebraic expressions. Then, you'll learn how to "solve for x" as well as for "x" and "y." "Solving for x" is one of the two basic algebraic skills that you should master. The other—and more difficult—skill involves creating or "setting up" an equation which you will then solve by isolating a variable on one side. We'll work on this in Chapters 11 and 12, where we'll set up and solve different kinds of equations for those dreaded word problems.

Algebraic Equations—Solving For X

Let's be clear as to what we're talking about here. *Algebra* involves the use of *variables* (such as x or a) in mathematical expressions to represent unknown values. An *algebraic expression* is any mathematical expression that includes one or more variables. Here is a simple example involving one variable: $x + 3$. Of course, "$x + 3$" doesn't mean much in itself, does it? That's because x could be any value; in other words, its value is variable (hence, the term "variable"). x must equal a particular value before it's of much use. This is why algebraic expressions are usually used to form equations, which set two expressions equal to each other. For example:

$$x + 3 = 1$$

Now we can determine the value of x. How? By isolating x on one side of the equation. To accomplish this, we need to get rid of that "+3" from the left side of the equation. To do so, subtract 3 from both the left and right sides of the equation:

$$x + 3 = 1$$
$$x + 3 - 3 = 1 - 3$$
$$x = -2$$

If you don't remember anything else from this chapter, remember this key to solving any equation:

> **Whatever operation you perform on one side of an equation you must also perform on the other side; otherwise, the two sides won't be equal.**

Depending on the particular equation with which you are dealing, you may need to perform one or more of the following operations *on both sides*:

1. addition or subtraction

2. multiplication or division

3. cross-multiplication

4. raising both sides to a power or finding a root of both sides

Performing any of these operations on both sides does not change the equality; it merely restates the equation in a different form. Let's take a look at each of these categories to see when and how to use each one.

1. Add or subtract the same term from both sides of the equation.

To solve for x, you may need to either add or subtract a term from both sides of an equation. Let's try an example that involves both addition and subtraction:

$$2x - 6 = x - 9$$

First, move the x from the right side of the equation to the left by subtracting it from both sides:

$$2x - 6 - x = x - 9 - x$$
$$x - 6 = -9$$

Next, isolate x by adding 6 to both sides:

$$x - 6 + 6 = -9 + 6$$
$$x = -3$$

2. Multiply or divide both sides of the equation by the same non-zero term.

To solve for x, you may need to either multiply or divide a term from both sides of an equation. Let's try an example that involves both multiplication and division:

$$-12 = \frac{11}{x} \qquad \text{original equation}$$

$$(-12)(x) = \left(\frac{11}{x}\right)(x) \qquad \text{multiply both sides by } x \text{ to remove } x \text{ from the denominator}$$

$$-12x = 11 \qquad \text{cancel } x \text{ in the numerator and denominator on the right side}$$

$$\frac{-12x}{-12} = \frac{11}{-12}$$ isolate x by dividing both sides by –12

$$x = -\frac{11}{12}$$ cancel –12 from the numerator and denominator on the left side

Let's try one that involves subtraction and division:

$$15x + \frac{1}{3} = 3$$ original equation

$$15x = 2\tfrac{2}{3} \text{ or } \tfrac{8}{3}$$ isolate the x-term by subtracting 1/3 from both sides

$$\frac{15x}{15} = \frac{\frac{8}{3}}{15}$$ isolate x by dividing both sides by 15

$$x = \left(\frac{8}{3}\right)\left(\frac{1}{15}\right)$$ multiply $\tfrac{8}{3}$ by the reciprocal of 15

$$x = \frac{8}{45}$$ combine the two fractions to obtain the value of x

3. **If each side of the equation is a fraction, your first step is to "cross-multiply."**

Where the original equation equates two fractions, use *cross-multiplication* to eliminate the fractions. Multiply the numerator from one side of the equation by the denominator from the other side. (In effect, cross-multiplication is a shortcut method of multiplying both sides of the equation by both denominators.) Set the product equal to the product of the other numerator and denominator:

$$\frac{7a}{8} \diagdown\kern-0.9em\diagup \frac{a+1}{3}$$ original equation (cross-multiplication indicated by arrows)

$$(3)(7a) = 8(a + 1)$$ result of cross-multiplication

$$21a = 8a + 8$$ distributing 8 to both a and 1

$$21a - 8a = 8a + 8 - 8a$$ isolating a-terms on one side by subtracting $8a$ from both sides

$$13a = 8$$ the a-terms are now isolated

$$\frac{13a}{13} = \frac{8}{13}$$ isolating a by dividing both sides by 13

$$a = \frac{8}{13}$$ cancel the "13"s on the left side

4. Squaring both sides of the equation to eliminate radical signs.

Where the variable is under a square-root radical sign, remove the radical sign by squaring both sides of the equation. (Use a similar technique for cube and other roots.)

$$\frac{11}{3}\sqrt{2x} = 2$$ original equation

$$\sqrt{2x} = (2)\left(\frac{3}{11}\right)$$ isolate the x-term

$$\left(\sqrt{2x}\right)^2 = \left(\frac{6}{11}\right)^2$$ combine terms, then square both sides of the equation

$$2x = \frac{36}{121}$$ both sides of the equation squared

$$\left(\frac{1}{2}\right)2x = \left(\frac{36}{121}\right)\left(\frac{1}{2}\right) \qquad \text{isolate } x$$

$$x = \frac{18}{121} \qquad \text{solution}$$

Warning: Be careful when you square both sides of an equation. In some instances, doing so will reveal *two* possible values for your variable. Here's an example:

$$6x = \sqrt{3x} \qquad \text{original equation}$$

$$36x^2 = 3x \qquad \text{square both sides of the equation}$$

$$36x^2 - 3x = 0 \qquad \text{isolate the } x\text{-terms on the left side}$$

$$x(36x - 3) = 0 \qquad \text{factor out an } x \text{ from each term of the left side}$$

$$x = 0 \text{ , } 36x - 3 = 0 \qquad \text{since the product of the terms "}x\text{" and "}36x - 3\text{" is 0, one of these two terms must equal 0.}$$

$$x = 0 \text{ , } \frac{1}{12} \qquad \text{solving for } x \text{ in the righthand equation gives us our two possible values for } x$$

How do you know if you're going to end up with two possible values for your variable? Well, when you encounter a variable with an exponent of 2, you're probably dealing with what's called a *quadratic equation*, which may hold as many as two possible solutions. Don't panic; quadratic equations will be examined in more detail later.

Manipulating Algebraic Expressions

Most algebra problems require you to manipulate algebraic expressions—that is, to restate them in some other form. Keep in mind that all of the rules that you learned for arithmetical operations and for exponents and roots apply to algebraic terms and expressions, too. You may wish to go back and review those rules before continuing here. Keep in mind the following four methods of manipulating algebraic expressions (this should be review):

1. Simplify a term by combining and/or canceling. Here are some examples:

$$(x^2)^3 = x^6$$ multiply exponents

$$\frac{a^4 b^3}{a^2 bc} = \frac{a^2 b^2}{c}$$ cancel a^2 and b from both numerator and denominator

$$(xy)\left(\frac{3y}{x}\right) = 3y^2$$ combine the y's and cancel x from both numerator and denominator

$$\sqrt{\frac{16x^2 y}{4x^3 y^2}} = \sqrt{\frac{4}{xy}} = \frac{2}{\sqrt{xy}} = \frac{2\sqrt{xy}}{xy}$$ cancel terms, remove "4" from the radical, then remove the radical from the denominator

2. Add or subtract terms having the same variable and same exponent. Here are some examples:

$$a^4 + a^4 = 2a^4$$ adding two terms with the same variable (with the same exponent)

$$\frac{1}{3}y^3 - 6y^3 = -\frac{16}{3}y^3$$ subtracting one term from another, where the terms include the same variable (with the same exponent)

$$2\sqrt{x^2 + y} + \sqrt{x^2 + y} = 3\sqrt{x^2 + y}$$ adding two terms which contain the same radical (and value thereof)

These expressions, however, cannot be combined:

$$a^2 + a$$ exponents are different

$$2a + 2b$$ variables are different

3. Factor out numbers and variables common to all terms; for example:

$$2x + 4xy + 10x^2y^2 =$$
$$2x(1 + 2y + 5xy^2)$$

each term includes coefficient 2 and variable x

4. Distribute a term among two or more other terms (the reverse of #3); for example:

$$-3b(-8x - bx + 3) =$$
$$24bx + 3b^2x - 9b$$

$-3b$ distributed among three other terms

Solving Algebraic Inequalities

Algebraic inequalities are solved in the same manner as equations. Isolate the variable on one side of the equation, factoring and canceling wherever possible. However, one important rule distinguishes inequalities from equations: Whenever you *multiply or divide by a negative number*, the order of the inequality—that is, the inequality symbol—must be *reversed*. Here's an example:

$$12 - 4x < 8$$

original inequality

$$-4x < -4$$

12 subtracted from each side; inequality unchanged

$$x > 1$$

both sides divided by −4; inequality reversed

Notice that the inequality remained unchanged when we subtracted the same number from each side.

Linear Equations with More than one Variable

What we've covered up to this point is pretty basic stuff. If you haven't quite caught on, you should probably stop here and consult a basic algebra workbook for more practice. On the other hand, if you're with me so far, let's forge ahead and add another variable to our equations! (Try to control your excitement.) Are you so allergic to algebra that you break out in a rash when confronted with *two* equations and *two* variables? Well, handling these problems is not any more difficult than what we've done so far in this chapter. So put away the calamine lotion, sharpen your pencil, and let's figure this out.

Once again, lets revisit "$x + 3 = 1$." Recall that we had no trouble at all solving for x by subtracting 3 from both sides of the equation; we determined that value of x is -2 and *only* -2. No other values of x will satisfy the equation. Try plugging in any other value for x; the left side doesn't equal the right side unless $x = -2$.

Now...let's change the equation by adding a second variable, y:

$$x + 3 = y + 1$$

Quick...what's the value of x? It depends on the value of y, doesn't it? Similarly, the value of y depends on the value of x. Without more information about either x or y, we're stuck; well, not completely. We can express x "in terms of y," and we can express y "in terms of x":

$$x = y - 2$$

$$y = x + 2$$

In fact, exam problems often call for you to do just that. Let's look at one more:

$$4x - 9 = \frac{3}{2}y$$

Solving for x in terms of y:

$$4x = \frac{3}{2}y + 9$$

$$x = \frac{3}{8}y + \frac{9}{4}$$

Solving for y in terms of x:

$$\frac{4x - 9}{\frac{3}{2}} = y$$

$$\frac{8}{3}x - 6 = y$$

Well, all of this "in terms of" stuff is nice to know, but it begs the real question: What are the values of x and y? Let's find out.

Solving for **X** and **Y**

Here again is our Catch 22 when confronted with *one* equation which includes *two* variables x and y: you can't know the value of x unless you know the value of y, but you can't know the value of y unless you know the value of x. Well...Equation #2 to the rescue! Here's a system of two equations with two variables:

Equation 1: $\frac{2}{5}x + y = 3y - 10$

Equation 2: $y = 10 - x$

There are two different methods for finding the values of two variables, given two equations: (1) the *substitution* method and (2) the *addition-subtraction* method. You can also combine these two approaches in a "hybrid" method. Let's determine the values for the two variables that satisfy both equations, using each method.

The Substitution Method

One way to solve for the two variables is to express one of them in terms of the other using one of the equations, and then *substitute* that value in the other equation. This equation can be solved and the value substituted in one if the equations to find the value of the other unknown. Here's one way to do it. The value of y in equation 2 can be substituted for y in equation 1 to solve for x (first combine the two y-terms in the first equation):

$$\frac{2}{5}x = 2y - 10$$

$$\frac{2}{5}x = 2(10 - x) - 10$$

$$\frac{2}{5}x = 20 - 2x - 10$$

$$\frac{2}{5}x = 10 - 2x$$

$$\frac{2}{5}x + 2x = 10$$

$$\frac{12}{5}x = 10$$

$$x = \frac{50}{12} \text{ or } \frac{25}{6}$$

This x-value can then be substituted for x in *either* equation 1 or 2 to determine the value of y. Let's use equation 2:

$$y = 10 - \frac{25}{6}$$

$$y = \frac{60}{6} - \frac{25}{6}$$

$$y = \frac{35}{6}$$

The Addition-Subtraction Method

Another way to solve for x and y is to make the coefficients of one of the variables the same (disregarding the sign) in both equations and either add the equations or subtract one equation from the other. Here's an example:

$$\text{Equation 1:} \quad 3x + 4y = -8$$

$$\text{Equation 2:} \quad x - 2y = \frac{1}{2}$$

In Equation 2, multiply both sides by 2 in order to solve for x by adding the equations:

$$
\begin{aligned}
3x + 4y &= -8 \\
2x - 4y &= \ \ 1 \\
\hline
5x + \ 0 \ &= -7 \\
x &= -\frac{7}{5}
\end{aligned}
$$

Similarly, to solve for y multiply both sides of Equation 2 by 3, then subtract Equation 2 from Equation 1:

$$
\begin{aligned}
3x + 4y &= -8 \\
3x - 6y &= \ \ \frac{3}{2} \\
\hline
0 + 10y &= -9\tfrac{1}{2} \\
10y &= -\frac{19}{2} \\
y &= -\frac{19}{20}
\end{aligned}
$$

The Hybrid Method

You can combine the addition-subtraction and substitution methods, solving for one variable using the former method, then substituting the value of that variable in either of the two equations to determine the value of the other variable. Here's how it works in the preceding example. Let's assume you've solved for x by adding the two equations but have not yet solved for y. You can plug x's value $\left(-\dfrac{7}{5}\right)$ into *either* equation to solve for y. You should get the same answer either way:

Equation 1	Equation 2
$3\left(-\dfrac{7}{5}\right) + 4y = -8$	$\left(-\dfrac{7}{5}\right) - 2y = \dfrac{1}{2}$
$-\dfrac{21}{5} + 4y = -8$	$-2y = \dfrac{1}{2} + \dfrac{7}{5}$
$4y = -\dfrac{40}{5} + \dfrac{21}{5}$	$-2y = \dfrac{19}{20}$
$4y = -\dfrac{19}{5}$	$-y = \dfrac{19}{20}$
$y = -\dfrac{19}{20}$	$y = -\dfrac{19}{20}$

If, instead, you've first solved for y by the addition-subtraction method, you can plug its value $\left(-\dfrac{19}{20}\right)$ into either equation to find the value of x. (Again, you'll get the same answer either way.)

Which Method Should You Use?

We've looked at three methods for handling two equations with two variables: addition-subtraction, substitution, and a hybrid method. Use the method with which you are most comfortable. In most cases, this will depend on the specific *numbers* (coefficients) appearing in the equations. For example, let's look one more time at these two equations:

$$3x + 4y = -8$$

$$x - 2y = \frac{1}{2}$$

Did you notice that changing Equation 2 to eliminate *either* x or y by addition-subtraction involved very easy calculations: multiplying both sides of the equation by either 3 or 2. We're dealing almost exclusively with integers using addition-subtraction. This system of equations was a nearly ideal candidate for addition-subtraction. The only way it could have been more ideal would be if no changes to either equation were necessary; for example, if the two original equations looked like this:

$$3x + 4y = -8$$

$$3x - 6y = \frac{3}{2}$$

To appreciate this point, go back and review the work we did to solve for x and y using all 3 methods. You should see that, in this case, addition-subtraction was the quickest and easiest method. However, the substitution method might be easier in other cases. Consider, for example, these two equations:

Equation #1: $\frac{7}{9}x + \frac{3}{8}y = \frac{11}{9}$

Equation #2: $\frac{\frac{9}{4}x - 18}{3} = 9y$

Quick...use addition-subtraction to solve for either x or y! Yikes! Not so quick and easy, is it? It would probably be easier in this case to isolate y on one side of the second equation, then substitute that expression for y in the first equation:

$$\frac{\frac{9}{4}x - 18}{27} = y \qquad \text{isolating } y \text{ in Equation \#2}$$

$$\frac{1}{12}x - \frac{2}{3} = y \qquad \text{simplifying Equation \#2}$$

$$\frac{7}{9}x + \frac{8}{3}\left(\frac{1}{12}x - \frac{2}{3}\right) = \frac{11}{9} \qquad \text{plugging } y\text{'s value into Equation \#1}$$

$$\frac{7}{9}x + \frac{2}{9}x - \frac{16}{9} = \frac{11}{9} \qquad \text{distributing terms}$$

$$x = \frac{27}{9} \text{ or } 3 \qquad \text{solving for } x$$

Remember: Use addition-subtraction if you can quickly and easily eliminate one of the variables. Otherwise, use substitution.

Equivalent Equations

In some cases, what appears to be a system of two equations with two variables is actually one equation expressed in two different ways. The following two equations are really the same equation and are said to be *equivalent* to each other:

$$a + b = 30$$

$$2b = 60 - 2a$$

Why? Although these two equations appear different, the second equation can be manipulated so that looks exactly the same as the first equation:

$$2b = 60 - 2a$$

$$2b = 2(30 - a)$$

$$b = 30 - a$$

$$a + b = 30$$

Solving for **X** When **X** is Squared

Up to this point in our quest for the value of x, we've been avoiding those pesky things called exponents. Well, now it's time to look x^2 "square in the face." To avoid muddying the waters, we'll leave second variables (like y) out of it and focus on equations with one variable raised to the power of two. To state it a bit more formally:

> **We will now examine equations that can be expressed in the general form: $ax^2 + bx + c = 0$, where a, b, and c are real numbers and $a \neq 0$.**

Note that the b-term and c-term are not essential—that is either b or c (or both) *can* equal zero. By the way, mathematicians call equations of this form "quadratic" equations ("quad" means "four," but x is squared; go figure.). Here are three examples:

Equation #1: $3x^2 = 10x$ (easier)

Equation #2: $3y = 4 - y^2$ (tougher)

Equation #3: $7z = 2z^2 - 15$ (toughest)

Because quadratic equations are non-linear, up to two distinct solutions (values of the variable x in the general form) *may* be possible. These solutions are called *roots*. A quadratic equation has *at most* two real-number solutions but may have only one or no real-number solutions. Any quadratic equation can be solved by using the following intimidating formula:

$$x = \frac{-b \pm \sqrt{b^2 - 4ac}}{2a} \quad \text{where a} \neq 0$$

However, on standardized exams you will not be required to use this formula. Most exam problems involving quadratic expressions can be *factored*, and thus can be simplified and solved without resorting to the formula above. Let's see how this is done.

Factoring Quadratic Expressions

Solving quadratic equations by factoring them calls for the following 3-step process:

1. Put the equation into the *standard form* ($ax^2 + bx + c = 0$).

2. *Factor* the terms of the left side of the equation into two linear expressions (roots) whose product is zero.

3. *Set each linear expression (root) equal to zero* and solve for the variable in each.

You will find that equations in which the *coefficient* of either the b-term or c-term (the number that precedes these variables) is *zero* are easier to factor, while equations with coefficients other than 0 and 1 are generally more difficult to factor. To illustrate this process, let's consider equations 1, 2, and 3 (from page 174):

Equation #1 ($3x^2 = 10x$)

In equation #1 (which does not include a *c*-term), it is easy to recognize that an x can be factored out as one of the two roots:

$$3x^2 = 10x \qquad \text{original equation}$$

$$3x^2 - 10x\,(+0) = 0 \qquad \text{general form } (ax^2 + bx + c = 0)$$

$$x(3x - 10) = 0 \qquad \text{the two roots are } x \text{ and } (3x - 10)$$

$$x = 0,\ 3x - 10 = 0 \qquad \text{each root is set equal to } 0$$

$$x = 0,\ \frac{10}{3} \qquad \text{two possible values of } x \text{ (two roots)}$$

Equation #2 ($3y = 4 - y^2$) and the FOIL method

Solving more complex quadratic equations is a bit trickier. For example, in equation #2 above, after putting the equation into the general form, you can see that there are no common variables or coefficients that can be factored out of all three terms:

$$3y = 4 - y^2 \qquad \text{original equation}$$

$$y^2 + 3y - 4 = 0 \qquad \text{general form } (ax^2 + bx + c = 0)$$

Instead, you must factor the quadratic expression into two linear *binomial* expressions, using what is called the *FOIL (First-Outer-Inner-Last)* method. Under the FOIL method, the sum of the following four terms is equivalent to the original (non-factored) quadratic expression):

(F) the product of the *first* terms of the two binomials

(O) the product of the *outer* terms of the two binomials

(I) the product of the *inner* terms of the two binomials

(L) the product of the *last (second)* terms of the two binomials

Note the following relationships:

F is the ax^2 term (first term) of the quadratic expression

O + I is the bx term (second term) of the quadratic expression

L is the c term (third term) of the quadratic expression

Applying the FOIL method to equation #2, we can first set up the following equation with two binomial factors:

$$(y + \,?)(y + \,?) = 0$$

$a x^2 + bx + c = 0$ *(handwritten, top margin)*

To determine the missing values of the two second terms, find two numbers the product of which is c (in this case, -4) and the sum of which is b (in this case, 3). Those two numbers are 4 and -1:

$$(y + 4)(y - 1) = 0$$

$$y + 4 = 0, \; y - 1 = 0$$

$$y = -4, \; 1$$

To check your work, reverse the process, using the FOIL method to multiply the two binomials together:

$$y^2 \text{ (first)} \; -y \text{ (outer)} + 4y \text{ (inner)} - 4 \text{ (last)} = 0$$

$$y^2 \, (-y + 4y) - 4 = 0$$

$$y^2 + 3y - 4 = 0$$

Equation #3 ($7z = 2z^2 - 15$) and the **FOIL** method

In equation #3, z^2 has a coefficient of 2. This complicates the process of factoring into two binomials. A bit of trial and error may be required to determine all coefficients in both binomials. Let's restate the equation in the general form and set up two binomial roots:

$$7z = 2z^2 - 15$$

$$2z^2 - 7z - 15 = 0$$

$$(2z + ?)(z + ?) = 0$$

(handwritten, right side)
$2z^2 - 7z + 15 = 0$
$(2z + \quad)(z + \quad)$

One of the two missing constants must be negative, since their product (the "L" term under the FOIL method) is -15. The possible integral pairs for these constants are: $(1,-15)$, $(-1,15)$, $(3,-5)$ and $(-3,5)$. Substituting each value pair for the two ?s in the equation above reveals that 3 and -15 are the missing constants (remember to take into account that the first x-term includes a coefficient of 2):

$$(2z + 3)(z - 5) = 0$$

Checking your work by reversing the process:

$$2z^2 - 10z + 3z - 15 = 0$$

$$2z^2 - 7z - 15 = 0$$

Now, solving for z:

$$(2z + 3)(z - 5) = 0$$

$$2z + 3 = 0, \ z - 5 = 0$$

$$z = -\frac{3}{2}, \ 5$$

Non-Linear Expressions with Two Variables

We need to add just one more wrinkle to the quadratic quilt before leaving the subject. On your exam you may also encounter non-linear expressions with *two* variables. If so, you probably won't have to determine numerical values for the variables. Instead, your task will be to factor and simplify the expression. Two related non-linear expressions that appear over and over again on the exams are worth noting here (use the FOIL method to verify these equations):

$$x^2 + y^2 = x^2 + 2xy + y^2$$

$$(x - y)^2 = x^2 - y^2$$

Memorize each of these two expressions in both factored and non-factored forms. When you see either form on the exam, in all likelihood the problem will require you to convert it to the other form.

Quiz Time

Here are 10 problems to test your skill in applying the concepts in this chapter. After attempting all 10 problems, read the explanations that follow. Then go back to the chapter and review your trouble spots. If you can handle the easier problems *and* the more challenging ones, consider yourself a "smart test taker"!

Easier

1. $\dfrac{2y}{9} = \dfrac{y-1}{3}$. What is the value of y?

2. $x + y = a$, and $x - y = b$. What is the value of x in terms of a and b?

3. $2\sqrt{x+1} - 4 = 8$. What is the value of x?

4. Express $x^2 - x = 20$ as the product of two binomials.

5. $-2x > -5$. Isolate x on one side of the inequality.

More Challenging

6. $x^2 - 4x = 21$. Find all possible values of x.

7. $x + y = 16$ and $x^2 - y^2 = 48$. What is the value of $x - y$?

8. $\dfrac{3x-1}{\dfrac{3}{x}} = 10$. Find all possible values of x.

9. $3x + 2y = 5a + b$ and $4x - 3y = a + 7b$. Express x in terms of a and b.

10. $a^2 - b^2 = 100$ and $a + b = 25$. What is the value of $a - b$?

Answers and Explanations

1. $y = 3$. Using cross-multiplication, here are the steps:

$$9(y - 1) = 2y(3)$$
$$9y - 9 = 6y$$
$$3y = 9$$
$$y = 3$$

2. $x = \frac{1}{2}(a + b)$. You can add the two equations, then isolate x:

$$x + y = a$$
$$\underline{x - y = b}$$
$$2x = a + b$$
$$x = \frac{1}{2}(a + b)$$

3. $x = 35$. Here are the steps:

$$2\sqrt{x + 1} = 12$$
$$\sqrt{x + 1} = 6$$
$$x + 1 = 36$$
$$x = 35$$

4. Use the FOIL method (in reverse) as follows:

$$x^2 - x - 20 =$$
$$x^2 - 5x + 4x - 20 =$$
$$(x - 5)(x + 4)$$

5. Multiply both sides of the equation by -1 and reverse the order of the inequality:

$$2x < 5$$
$$x < \frac{5}{2}$$

6. $x = 7$ or -3. Here are the steps:

$$x^2 - 4x - 21 = 0$$
$$(x - 7)(x + 3) = 0$$
$$x - 7 = 0 \quad x + 3 = 0$$
$$x = 7 \text{ or } -3$$

7. $x - y = 3$. Here are the steps:

$$x^2 - y^2 = (x + y)(x - y) = 48$$

Substituting 16 for $x + y$:

$$16(x - y) = 48$$
$$x - y = 3$$

8. $x = \dfrac{10}{3}$ or -3. Invert the denominator fraction and multiply it by the numerator:

$$\frac{3x^2 - x}{3} = 10$$
$$3x^2 - x = 30$$
$$3x^2 - x - 30 = 0$$
$$(3x - 10)(x + 3) = 0$$
$$x = \frac{10}{3} \text{ or } -3$$

9. $x = a + b$. Multiply the first equation by 3, the second by 2, then add:

$$9x + 6y = 15a + 3b$$
$$\underline{8x - 6y = 2a + 14b}$$
$$17x = 17a + 17b$$
$$x = a + b$$

10. $a - b = 4$

$a^2 - b^2 = (a+b)(a-b) = 100$

Substituting 25 for $a + b$:

$25(a - b) = 100$

$a - b = 4$

11

Statistics, Sets,
and Sequences

In this chapter we'll look at numbers as statisticians do—we'll evaluate groups, or *sets*, of numbers, looking for patterns and other insights about those numbers. Along the way, you'll have the occasional opportunity to flex your recently-developed algebraic muscles.

Wait! Don't buy that one-way plane ticket to Las Vegas with the idea that you're going to learn enough about probability in this chapter so that you can "beat the house" every time in blackjack or poker. The area of probability is one of many complex aspects of the specialized field of statistics that is beyond the scope of basic math tests (and beyond the scope of this book). So you won't become an overnight millionaire just by reading this chapter. The good news, however, is that **basic math exams cover** only these elementary aspects of statistics:

- arithmetic mean (simple average) and "weighted" average

- median and mode

- sets (combinations)

- overlapping groups

- sequential patterns among numbers (or variables) in a set

Some of these topics involve logic more than they do math. Nevertheless, they're covered on the exams, so we'll cover them here.

Arithmetic Mean (Simple Average)

We all use the concept of *simple average* every day:

- How long does it take you to get to school…on average?
- How much is your average weekly grocery bill?
- What is your favorite baseball player's batting average?

Have you ever stopped to think about exactly how you compute averages such as these? Here's what's happening: you're adding up all the numbers and dividing the sum by the number of numbers. Let's state the formula a bit more formally:

The *arithmetic mean*, or *simple average*, of a set of numbers or other terms is determined by adding the terms together and dividing the resulting sum by the number of terms in the set.

For example, letting *A* equal the average of five terms—*v*, *w*, *x*, *y* and *z*. *A* can be expressed as follows:

$$A = \frac{v + w + x + y + z}{5}$$

Here's a numerical example. Let's say that during 4 successive weeks, your weekly grocery bills are as follows: $32.50, $27.00, $39.20, $26.10. To determine your average weekly grocery bill, add these four amounts together and divide the sum by 4:

$$\$32.50 + \$27.00 + \$39.20 + \$26.10 = \$124.80$$

$$\frac{\$124.80}{4} = \$31.20$$

Averages can be deceiving. An average number is not necessarily midway in value between the highest and lowest numbers. Nor does it have to be close in value to any of the numbers in the set. For example, if you catch an average of 10 fish per day during a 7-day fishing trip, it's quite possible you caught 70 on the first day and spent the next 6 days in your cabin reading a good book:

$$70 + 0 + 0 + 0 + 0 + 0 + 0 = 70$$

$$\frac{70}{7} = 10$$

You caught 10 fish per day, on average, although on none of the seven days did you catch anywhere near 10 fish.

Weighted Average

The concept of simple average (arithmetic mean) is just that--simple. Not-so-simple, however, is the concept of "weighted average." When some numbers among the terms to be averaged are given greater "weight" than others, however, our simple-average formula is inadequate, because the various terms must be adjusted to reflect differing weights. As a simple illustration, suppose a student receives a grades of 80 and 90 on two exams, but the former grade receives three times the weight of the latter exam. The student's weighted-average grade is not 85 but rather some number closer to 80 than 90. One way to approach this problem is to think of the first grade (80) as three scores of 80, which added to the score of 90 and divided by 4 (not 2) results in the weighted average:

$$WA = \frac{80 + 80 + 80 + 90}{4} = \frac{330}{4} = 82.5$$

You can also approach this problem more intuitively (less formally). You are looking for a number between 80 and 90 (a range of 10). The simple average would obviously lie midway between the two. Given that the score of 80 receives three times the weight of the score of 90, the weighted average is three times closer to 80 than to 90, or $\frac{3}{4}$ of the way from 90 to 80. Dividing the range into four segments, it is clear that the weighted average is 82.5. Similarly, if 80 received twice the weight of 90, the weighted average would be $83\frac{1}{3}$, and if 80 received 4 times the weight of 90, the weighted average would be 82.

Let's apply both the formal algebraic and less formal approaches to two examples:

First example (weighted average is given)

Q: Mike's average monthly salary for the first four months was $3,000. What must his average monthly salary be for each of the next 8 months, so that his average monthly salary for the year is $3,500?

In this relatively easy example, the $3,000 salary receives a weight of 4, while the unknown salary receives a weight of 8. You can approach this problem in a strict algebraic manner as follows:

$$3,500 = \frac{4(3,000) + 8x}{12}$$
$$(12)(3,500) = 12,000 + 8x$$
$$30,000 = 8x$$
$$x = 3,750$$

Mike's salary for each of the next 8 months must be $3,750 in order for Mike to earn an average of $3,500 a month during the entire twelve months. You can also approach this problem more intuitively. $\frac{1}{3}$ of the monthly salary payments are "underweighted" (less than the desired $3,500 average) by $500. Thus, to achieve the desired average with twelve salary payments, the remaining $\frac{2}{3}$ of the payments must be overweighted (exceed $3,500) by half that amount—i.e., by $250.

Second example (terms to be averaged are given)

Q: Cynthia drove for 7 hours at an average rate of 50 miles per hour and for 1 hour at an average rate of 60 miles per hour. What was her average rate for the entire trip?

As in the exam-grade illustration earlier, think of Cynthia's average rate as the average of 8 equally-weighted 1-hour trips. 7 of those trips receive a weight of 50, and one of the trips receives a weight of 60. Expressed algebraically:

$$AR = \frac{7(50) + 60}{8} = \frac{350 + 60}{8} = \frac{410}{8} = 51.25$$

Cynthia's average rate during the entire trip was 51.25 miles per hour. Of course, you can approach this problem more intuitively. The single faster hour of the 8-hour trip boosts what would otherwise have been a 50-m.p.h. average rate up $\frac{1}{8}$ of the way to 60—i.e., up by 1.25 to 51.25.

Median and Mode

No discussion about arithmetic mean would be complete without at least mentioning the concepts of *median* and *mode*. In a set of numbers ordered from least to greatest, the *median* is defined as the middle value (if the series includes an *odd* number of terms) or the average of the two middle values (if the series includes an *even* number of terms). Thus, the median of five terms—a, b, c, d, and e (in order of value)—is c, while the median of four terms—a, b, c, d (in order of value)—is $\frac{b+c}{2}$. Here's an example:

$$\{3, -2,\ 10, 4, -11, -7\} \quad \text{set of numbers}$$

$$\{-7, -2, 3, 4, 10, 11\} \quad \text{rearranged in order of value}$$

$$3.5 \quad \text{median: } \frac{3+4}{2}$$

Don't confuse mean with median. For the same set of values, the mean (average) and the median can be, but are not necessarily, the same. For example:

The set $\{3,4,5,6,7\}$ has both a mean and median of 5.

The set $\{-2,0,5,8,9\}$ has a mean of 4 but a median of 5.

The *mode* is simply the number that appears most often among a set of numbers. Here's an example involving a set of six numbers:

$$\{5, 6, 4, 6, 3, 7\}$$

The mode is 6.

If two numbers appear equally as often, the set is said to be *bimodal*. Nothing too interesting (or testworthy) here, is there? So let's move on.

Set Combinations

A *set* involves a group of two or more numbers or other terms. An exam question might ask you to determine the number of *possible combinations* of terms within the group or among terms in two groups. For example, consider these two sets, aptly called Set 1 and Set 2, and the question that follows:

Set 1: {a, b, c, d, e}

Set 2: {x, y, z}

Q: How many different sets of five members are possible by combining three members from Set 1 with two members from Set 2?

In formal mathematics (more specifically, the field of probability) such problems generally call for the use of *factorials*. On standardized exams, however, problems such as this one are usually simple enough to analyze less formally and more intuitively—without resort to mathematics at all (except to add the number of possibilities).

There isn't anything conceptually difficult about the question above. It's simply a matter of being methodical and careful as you tally up all possibilities. Here's a method that you should apply to any problem like this one:

1. Combine terms as far to the left as possible before moving to the right. In the question above, for example, first consider all possible combinations with *a*:

{a b c}	{a c d}
{a b d}	{a c e}
{a b e}	{a d e}

2. Consider all combinations including *b*, aside from those already accounted for (in other words, move to the *right* only):

$$\{b\ c\ d\}$$
$$\{b\ c\ e\}$$
$$\{b\ d\ e\}$$

3. Consider all combinations including c, aside from those already accounted for (in other words, move to the *right* only):

$$\{c\ d\ e\}$$

All three-member combinations including d and e have already been accounted for. Thus, there are a total of 10 possible three-member combinations from Set 1.

4. Similarly, from Set 2 there are a total of 3 possible two-member combinations:

$$\{x\ y\}$$
$$\{x\ z\}$$
$$\{y\ z\}$$

Thus, the total number of combinations asked for by the question is 10×3, or 30.

Overlapping Groups

Don't confuse problems involving *overlapping groups* with the type of problem we just looked at, which involves combinations of set members. In overlapping group problems, we're dealing with different groups (sets) that have *certain members in common*. These problems come in one of two varieties on standardized tests: *single overlap* and *double overlap*. Let's look at an example of each.

Example of a "single overlap" problem

Q: Each of the 24 people auditioning for a community-theater production is either an actor, a musician, or both. If 10 of the people auditioning are actors and 19 of the people auditioning are musicians, how many of the people auditioning are musicians but not actors?

This problem presents three mutually-exclusive sets: (1) actors who are not musicians, (2) musicians who are not actors, and (3) actors who are also musicians. The total number of people among these three sets is 24. You can represent this scenario with the following algebraic equation (n = number of actors / musicians), solving for $19 - n$ to respond to the question:

$$(10 - n) + n + (19 - n) = 24$$
$$29 - n = 24$$
$$n = 5$$
$$19 - n = 14$$

There are 14 musicians auditioning who are not actors. With problems such as this one, it may be helpful to use a *Venn Diagram* in which overlapping circles represent the set of musicians and the set of actors:

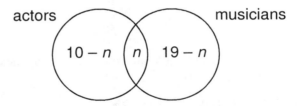

You can, of course, approach this problem less formally as well: The number of actors plus the number of musicians equals 29 (10 + 19 = 29); however, only 24 people are auditioning. Thus, 5 of the 24 are actor-musicians, so 14 of the 19 musicians must not be actors.

Example of a "double overlap" problem

Q: Adrian owns 48 neckties, each of which is either 100% silk or 100% polyester. 40% of his ties are striped, and 13 of his ties are silk. How many ties does Adrian own that are polyester but are not striped?

This double overlap problem involves four distinct sets: (1) striped silk ties, (2) striped polyester ties, (3) non-striped silk ties, and (4) non-striped polyester ties. The number of ties among these four discreet sets totals 48. Although you can approach this problem formally, another way to handle it is to use a table to represent the four sets. Fill in the information given in the problem (the value the question asks for is indicated by a "?"):

	silk	polyester	totals
striped			40%
non-striped		?	60%
totals	13	35	

Given that 13 ties are silk (left column), 35 ties must be polyester (right column). Also, given that 40% of the ties are striped (top row), 60% must be non-striped (bottom row). Thus, 60% of 35 ties, or 19 ties, are polyester and non-striped.

Sequential Patterns

What if an exam question were to call for you to determine the sum of a long string of numbers such as this one:

$$\{-17, -15, 12, -6, -3, 0, 3, 6, 12, 15, 18, 21\}$$

You could determine the sum by performing time-consuming addition; however, keep in mind that there is probably a quicker way. Remember: One of the skills measured on standardized math tests is your ability to recognize quick, intuitive ways to solve the problems. Notice that all positive numbers, with the exception of 18 and 21, have a negative-number counterpart. All of these numbers add up to 0 and can be ignored in computing the sum of the digits. Only the numbers 18 and 21, as well as the negative number -17, need by considered. 18 and -17 "average out" to 1. Thus, the sum of all of the digits is $21 + 1$, or 22.

A real exam question is more likely to add a wrinkle of complexity to the type of problem presented above. A problem might, for instance, ask you to compare the sum of two *sets* of digits—for example, the difference between the sum of the integers in these two sets:

Set 1: {all integers between and including 14 and 41}

Set 2: {all integers between and including 11 and 40}

Instead of computing both sums, cancel out the integers 14 through 40 in each set; these are the integers that are common to both sets. That leaves only the integer 41 in Set 1 and the integers 11, 12, and 13 in Set 2. So the difference between the sum of the integers in the two sets is 5:

$$\{Set\ 1\} - \{Set\ 2\} = 41 - (11 + 12 + 13) = 41 - 36 = 5$$

Let's try one more, adding another wrinkle to the basic problem. What's the difference between the sum of the numbers in each of these two sets:

Set 1: {all odd integers between 0 and 60}

Set 2: {all even integers between 0 and 61}

Because Set 1 includes only *odd* integers, while Set 2 includes only *even* integers, there's no overlap here—no common members that we can cancel. Are we doomed? Certainly not. Consider as a "pair" the integer 1 from Set 1 and the integer 2 from Set 2. Their difference is 1. Now consider the next integer from each set—3 and 4. Again, the value of the integer in Set 2 is one more than the value of the corresponding integer in Set 1. Each set includes a total of 30 integers. Thus, repeat this analysis 30 times (for every integer in each set) for a total difference of 30. No time-consuming computations required!

Quiz Time

Here are 10 problems to test your skill in applying the concepts in this chapter. After attempting all 10 problems, read the explanations that follow. Then go back to the chapter and review your trouble spots. If you can handle the easier problems *and* the more challenging ones, consider yourself a "smart test taker"!

Easier

1. Determine both the *arithmetic mean* and *median* of the following set of numbers: {−2, 7, 22, 19}.

2. ABC Company pays an average of $140 per vehicle each month in outdoor parking fees for 3 of its 8 vehicles. The company pays garage parking fees for the remaining 5 vehicles. If ABC pays an average of $240 per vehicle overall each month for parking, how much does ABC pay per month in *garage* parking fees for its vehicles?

3. Every customer at a particular coffee house likes either cream or sugar, or both, in their coffee. 48% of the customers like both cream and sugar, while 22% like sugar but not cream. What percentage of the customers like cream but not sugar?

4. Two competitors battle each other in each match of a tournament among 9 competitors. What is the minimum number of matches that must occur in order for every competitor to battle every other competitor?

5. What is the difference between the sum of the integers 15 through 33, inclusive, and the sum of the integers 11 through 31, inclusive?

More Challenging

6. In a group of m workers, if b workers earn D dollars per week and the rest earn half that amount each, express the total number of dollars paid to the entire group of workers in a week in terms of D, b, and m?

7. Among all sales staff at Listco Corporation, college graduates and those without college degrees are equally represented. Each sales staff member is either a level-1 or level-2 employee. Level-1 college graduates account for 15% of Listco's sales staff. Listco employs 72 level-1 employees, 30 of whom are college graduates. How many sales staff members without college degrees are level-2 employees?

8. During the first 3 weeks of a 10-week diet program, Bob lost an average of 5 pounds per week. During the final 7 weeks of the program, he lost an average of 2 pounds per week. If he lost the same amount of weight during the first 3 weeks of the program as during the last 3 weeks of the program, how much weight had Bob lost after the seventh week of the diet program?

9. In an election between two candidates—Lange and Sobel— 70% of the voters voted for Sobel. 60% of the voters in the election were male. If 35% of the female voters voted for Lange, what percentage of the male voters voted for Sobel?

10. At a particular ice-cream parlor, customers can choose among five different ice-cream flavors and may choose either a sugar cone or a waffle cone. Considering both ice-cream flavor and cone type, how many distinct triple-scoop cones with three different flavors are available?

Answers and Explanations

1. The arithmetic mean is 11.5. The median is 13.

$$\frac{-2+7+22+19}{4} = \frac{46}{4} = 11.5 \text{ (arithmetic mean)}$$

$$\frac{7+19}{2} = 13 \quad \text{(median)}$$

2. The total parking fee paid by ABC each month is $1,920 ($240 × 8). Of that amount, $420 is paid for outdoor parking for three cars. The difference ($1,920 − $420 = $1,500) is the total garage parking fee paid for the other 5 cars.

$$3(140) = 420$$
$$5(300) = 1500$$
$$420 + 1500 = 1920$$
$$1920 \div 8 = 240$$

3. The answer is 30%. The first sentence of the problem identifies three distinct groups, which add up to 100% of the customers. Two of the groups account for 70% (48% + 22%) of the customers, the third group must account for the remaining 30%.

4. The answer is 36. Competitor #1 must engage in 8 matches. Competitor #2 must engage in 7 matches not already accounted for (the match between #1 and #2 has already been tabulated). Similarly, competitor #3 must engage in 6 matches other than those accounted for, and so on. The minimum number of total matches = 8 + 7 + 6 + 5 + 4 + 3 + 2 + 1 = 36.

5. The answer is 15. It is unnecessary to add all the terms of each sequence. Instead, notice that the two sequences have in common integers 15 through 31, inclusive. Thus, those terms cancel out, leaving 32 + 33 = 65 in the first sequence and 11 + 12 + 13 + 14 = 50 in the second sequence. The difference is 15.

6. The answer is $\frac{1}{2}D(b + m)$: The money earned by b workers at D dollars per week is bD dollars. The number of workers remaining is $(m - b)$, and since they earn $\frac{1}{2}D$ dollars per week, the money they earn is $\frac{1}{2}D(m - b) = \frac{1}{2}mD - \frac{1}{2}bD$. Thus, the total amount earned is $bD + \frac{1}{2}mD - \frac{1}{2}bd = \frac{1}{2}bd + \frac{1}{2}mD = \frac{1}{2}D(b + m)$.

7. The answer is 58. The information in this problem can be organized in a table:

	Level 1	Level 2	totals
CG	30 (15%)	70 (35%)	50%
non–CG	42 (21%)	58 (29%)	50%
totals	72 (36%)	128 (64%)	

8. The answer is 14 pounds. Given that Bob lost an average of 5 pounds per week during the first 3 weeks, his total weight loss during that period was 15 pounds. This is also the amount of weight he lost during the last three weeks. Thus, his total weight loss during all but the fourth through seventh weeks was 30 pounds. Given that he lost 29 pounds altogether during the 10 weeks $[(3 \times 5) + (7 \times 2)]$, he must have gained 1 pound during the fourth through seventh week. Accordingly, he had lost 14 pounds $(-15 + 1)$ after the first seven weeks.

9. The answer is 44%. Since 35% of 40% of the voters (female voters) voted for Lange, 14% $(.40 \times .35)$ of all voters were females who voted for Lange. The information in this problem can be organized in the following table (the total of all four percentages must be 100%):

	male	*female*	*totals*
Lange	16%	14%	(30%)
Sobol	44%	26%	(70%)
totals	(60%)	(40%)	

10. The answer is 20. Let {A, B, C, D, E} represent the set of ice-cream flavors. 10 triple-scoop combinations are available: {ABC}, {ABD}, {ABE}, {ACD}, {ACE}, {ADE}, {BCD}, {BCE}, {BDE}, and {CDE}. Each of these combinations is available on either of the two cone types. Thus, the total number of distinct ice-cream cones is 20.

12

Algebra
Word Problems

Word problems are those widely dreaded algebra (and geometry) questions set in a real life context. (don't take the term "real life" too literally) rather than in a pure mathematical setting. Test-takers who consider themselves "pretty good" at algebra often find word problems to be their Waterloo (that means downfall—Napoleon lost the Battle of Waterloo in a big way.) Take heart: Chapter 12 is here to help!

All word problems involve solving equations using the rules and techniques that you learned in Chapter 10. What makes word problems a bit trickier, though, is that you have to create—or "set up"— the equation(s) that you intend to solve. And to do this, you need to learn some formulas. That's the bad news. The good news is that you don't need to learn very many of them! The scholastic gods that mastermind standardized exams really don't have much imagination when it comes to word problems. They tend to fall back on the **tried-and-true categories** listed below and on the top of page 198 (the first two we've already explored in Chapter 11):

- weighted average problems (chapter 11)
- overlapping set problems (chapter 11)

- currency (coin and bill) problems
- work problems
- motion problems
- mixture problems
- age problems
- investment problems

In this chapter, we'll look at each one of these types (except for the ones we covered in Chapter 11). For each problem type, we'll look at one or more typical examples, and we'll learn how to apply the right formula to the solve the problem.

Currency (Coin and Bill) Problems

Currency problems are really quasi-weighted-average problems, since each item (bill or coin) in a problem is weighted according to it's monetary value. Unlike weighted average problems, however, the "average" value of all of the bills or coins is not at issue. In solving currency problems, remember:

1. You must formulate algebraic expressions involving both *number* of items (bills or coins) and *value* of items.

2. You should convert the value of all moneys to a common unit (e.g., cents or dollars) before formulating an equation. If converting to cents, for example, the number of nickels must be multiplied by 5, dimes by 10, and so forth.

Example: Coins of Different Value

Q: Jim has $2.05 in dimes and quarters. If he has four fewer dimes than quarters, how much money does he have in dimes?

Letting x equal the number of dimes, $x + 4$ represents the number of quarters. The total value of the dimes (in cents) is $10x$, and the total value of the quarters (in cents) is $25(x + 4)$ or $25x + 100$. Given that Jim has $2.05, the following equation emerges:

$$10x + 25x + 100 = 205$$

$$35x = 105$$

$$x = 3$$

Jim has 3 dimes; thus, he has 30 cents in dimes.

Work Problems

Work problems involve one or more workers (people or machines) accomplishing a task or job. In work problems, there is an inverse relationship between the number of workers and the time it takes to complete the job—in other words, the more workers, the quicker the job gets done. A work problem may specify the rates at which certain workers work alone and ask you to determine the rate at which they work together, or vice versa. Here is the basic formula for solving a work problem:

$$\frac{1}{x} + \frac{1}{y} = \frac{1}{A}$$

In this formula, x and y represent the time needed for each of two workers—x and y—to complete the job alone, and A represents the time it takes for both x and y to complete the job working (together). The reasoning is that in 1 unit of time (e.g., one hour) x performs $\frac{1}{x}$ of the job, y performs $\frac{1}{y}$ of the job, and x and y perform $\frac{1}{A}$ of the job.

In the real world, if two workers can perform a given task in the same amount of time working alone, it may not be possible for them to perform that same task in half that time working together. However, in work problems on math exams, you can assume that there is no individual efficiency gained or lost by two or more workers working together.

Let's look at two work problems, one requiring you to determine the aggregate rate of the workers (working together), the other requiring you to determine an individual worker's rate (working alone).

Example: Individual Rates Given

Q: One printing press can print a daily newspaper in 12 hours, while another press can print it in 18 hours. How long will the job take if both presses work simultaneously?

The rate of the faster press is $\frac{1}{12}$ (it can print $\frac{1}{12}$ of the paper in one hour), and the rate of the slower press is $\frac{1}{18}$:

$$\frac{1}{12} + \frac{1}{18} = \frac{1}{A}$$

$$\frac{3}{36} + \frac{2}{36} = \frac{1}{A}$$

$$\frac{5}{36} = \frac{1}{A}$$

$$5A = 36$$

$$A = \frac{36}{5}$$

It takes both presses $\frac{36}{5}$ hours, or 7 hours, 12 minutes, to print the daily paper working together.

Example: Aggregate Rate Given

Q: Peter and Belinda can make a particular quilt in two days when working together. If Peter requires 6 days to make the quilt alone, how many days does Belinda need make the quilt alone?

Peter can complete $\frac{1}{6}$ of the quilt in one day. The aggregate rate of Belinda and Peter working together is $\frac{1}{2}$ (together they can complete $\frac{1}{2}$ of the quilt in one day):

$$\frac{1}{6} + \frac{1}{b} = \frac{1}{2}$$

$$\frac{b+6}{6b} = \frac{1}{2}$$

$$2(b+6) = 6b$$

$$2b + 12 = 6b$$

$$4b = 12$$

$$b = 3$$

It takes Belinda 3 days working alone to make the quilt.

In some cases, a second "worker" may slow or impede the progress of the other worker, contributing a "negative" rate of work. Nevertheless, your approach should be basically the same, as the following example illustrates.

Example: Negative Rate of Work

Q: A certain water tank holds a maximum of 450 cubic meters of water. If a hose can fill the tank at a rate of 6 cubic meters per minute, but the tank has a hole through which a constant one-half cubic meters of water escapes each minute, how long will it take to fill the tank to its maximum capacity?

In this problem, the hole is the second "worker," but because it is acting counter-productively, its rate must be subtracted from the hose's rate to determine the aggregate rate of the hose and the hole. The hose alone would take 90 minutes to fill the tank. The hole alone would empty a full tank in 900 minutes. Thus, "working" together:

450

$$\frac{1}{90} - \frac{1}{900} = \frac{1}{A}$$

$$\frac{10}{900} - \frac{1}{900} = \frac{1}{A}$$

$$\frac{9}{900} = \frac{1}{A}$$

$$9A = 900$$

$$A = 100$$

It would take 100 minutes to fill the tank to its maximum capacity.

Motion Problems

Motion problems involve the linear movement of persons or objects over time. Fundamental to all motion problems is the following simple and familiar formula:

distance = rate × time   *(or)* $d = rt$

Some motion problems track two objects (or persons) which move either in the same direction or in opposite directions. Others involve one moving object (or person), tracking two parts or "legs" of a trip (e.g., away and back during a "round trip"). In any case, one of the three variables—distance, rate, or time—is *constant* (i.e., the same for both moving objects or both legs of a trip). This feature enables you to set up an equation and to solve for the missing value.

Nearly every motion problem falls into one of three categories:

1. two objects moving in opposite directions

2. two objects moving in the same direction

3. one object making a round trip

A fourth type of "motion" problem involves perpendicular (right-angle) motion—for example, where one object moves in a northerly direction while another moves in an easterly direction. However, this type is really just as much a geometry as an algebra problem, since the distance between the two objects is determined by applying the Pythagorean Theorem to determine the length of a triangle's hypotenuse. (We'll triangulate in Chapter 13.)

Let's take a look at one example of each of the three motion problems listed above.

Example: Motion in Opposite Directions (Time Constant)

Q: A passenger train and a freight train leave at 10:30 a.m. from stations which are 405 miles apart and travel toward each other. The rate of the passenger train is 45 miles per hour faster than that of the freight train. If they pass each other at 1:30 p.m., how fast was the passenger train traveling?

Notice in this problem that each train traveled exactly 3 hours—in other words, time is the constant in this problem. Let x equal the rate (speed) of the freight train. The rate of the passenger train can be expressed as $x + 45$. Substitute these values for time and rate into the motion formula for each train:

$$\text{freight}: (x)(3) = 3x$$

$$\text{passenger}: (x + 45)(3) = 3x + 135$$

The total distance covered by the two trains is given as 405 miles. Express this algebraically and solve for x:

$$(3x + 135) + (3x) = 405$$

$$6x = 270$$

$$x = 45$$

Accordingly, the rate of the passenger train was $45 + 45$ or 90 m.p.h.

Example: Motion in Same Direction (Distance Constant)

Q: Janice left her home at 11 a.m., traveling along Route 1 at 30 miles per hour. At 1 p.m., her brother Richard left home and started after her on the same road at 45 miles per hour. At what time did Richard catch up to Janice?

Notice that the distance covered by Janice is equal to that of Richard—i.e., distance is constant. Letting x equal Janice's time, Richard's time can be expressed as $x - 2$. Substitute these values for time and the values for rate given in the problem into the motion formula for Richard and Janice:

$$\text{Janice: } (30)(x) = 30x$$
$$\text{Richard: } (45)(x - 2) = 45x - 90$$

Because the distance is constant, Janice's distance as expressed algebraically above can be equated with Richard's distance, and the value of x can be determined:

$$30x = 45x - 90$$
$$15x = 90$$
$$x = 6$$

Janice had traveled 6 hours when Richard caught up with her. Because Janice left at 11:00 a.m., Richard caught up with her at 5:00 p.m.

Example: Motion Involving Round Trip (Distance Constant)

Q: How far can Scott drive into the country if he drives out at 40 miles per hour, returns over the same road at 30 miles per hour and spends 8 hours away from home including a one-hour stop for lunch?

Scott's actual driving time is 7 hours, which must be divided into two parts: his time spent driving into the country and his time spent returning. Letting the trip out equal x, the return time is what remains of the 7 hours, or $7 - x$. Substituting these expressions into the motion formula for each of the two parts of Scott's journey:

$$going: \ (40)(x) = 40x$$

$$return: \ (30)(7 - x) = 210 - 30x$$

Because the journey is round trip, the distance going equals the distance returning. Accordingly, the value of x can be determined algebraically:

$$40x = 210 - 30x$$

$$70x = 210$$

$$x = 3$$

If Scott traveled 40 miles per hour for 3 hours, he traveled 120 miles.

Mixture Problems

In *mixture* problems, substances with different characteristics are combined, resulting in a particular mixture or proportion. There are really two types of mixture problems: wet and dry mixture:

Wet mixture problems involve liquids, gases, or granules, which are measured and mixed by volume or weight, not by number (quantity).

Dry mixture problems involve a number of discreet objects, such as coins, cookies, or marbles, that are measured and mixed by number (quantity) as well as by relative weight, size, value, etc.

Your approach toward wet and dry mixture problems should be similar. On the following pages, we'll take a look at an example of each type.

Example: Wet Mixture

Q: How many quarts of pure alcohol must be added to 15 quarts of a solution which is 40% alcohol to strengthen it to a solution which is 50% alcohol?

The original amount of alcohol is 40% of 15. Letting x equal the number of quarts of alcohol that must be added to result in a 50% alcohol solution, $.4(15) + x$ equals the amount of alcohol in the solution after adding more alcohol. This amount can be expressed as 50% of $(15 - x)$. Thus, the mixture can be expressed algebraically as follows:

$$(.4)(15) + x = (.5)(15 + x)$$

$$6 + x = 7.5 + .5x$$

$$.5x = 1.5$$

$$x = 3$$

3 quarts of alcohol must be added to result in a 50% alcohol solution.

If you have difficulty expressing mixture problems algebraically, use a table such as the one below to indicate amounts and percentages, letting x equal the amount or percentage you are asked to solve for:

	# of quarts ×	% alcohol =	amount of alcohol
original	15	40%	6
added	x	100%	x
new	$15 + x$	50%	$.5(15 + x)$

Example: Dry Mixture

Q: How many pounds of nuts selling for 70 cents per pound must be mixed with 30 pounds of nuts selling at 90 cents per pound to make a mixture which sells for 85 cents per pound?

The cost (in cents) of the nuts selling for 70 cents per pound can be expressed as $70x$, letting x equal the number we are asked to determine. This cost is then added to the cost of the more expensive nuts $[(30)(90) = 2,700]$ to obtain the total cost of the mixture, which can be expressed as $85(x + 30)$. Stated algebraically and solving for x:

$70x + (90)30 =$

$$70x + 2700 = 85(x + 30)$$

$$70x + 2700 = 85x + 2550$$

$$150 = 15x$$

$$x = 10$$

10 pounds of 70-cent-per-pound nuts must be added to make a mixture which sells for 85 cents per pound.

As with wet mixture problems, if you have trouble formulating an algebraic equation needed to solve the problem, indicate the quantities and values in a table such as the one below, letting x equal the value you are asked to determine:

	# of pounds	×	$ per pound	=	total value
cheaper	x		70		$70x$
expensive	30		90		2,700
mixture	$x + 30$		85		$85(x + 30)$

Age Problems

In *age* problems, you are asked to compare ages of two or more people at different points in time. You may be called upon to represent a person's age at the present time, several years from now, or several years ago. Any age problem will allow you to set up an equation to relate the ages of two (or more) people. For example:

- If x is 10 years younger than y at the present time, the relationship between x's age and y's age can be expressed as: $x = y - 10$ (or $x + 10 = y$).

- If a was twice as old as b five years ago, the relationship between their ages can be expressed as $2(a - 5) = b - 5$, where a and b are the present ages of a and b, respectively.

$\mathcal{E} = 24 + X$

Example: Age Problem

Q: Eva is 24 years older than her son Frank. In 8 years, Eva will be twice as old as Frank will be then. How old is Eva now?

Letting x equal Frank's present age, his age in 8 years can be expressed as $x + 8$. Similarly, Eva's present age can be expressed as $x + 24$, and her age in 8 years can be expressed as $x + 32$. Set up the following equation relating Eva's age and Frank's age eight years from now:

$$x + 32 = 2(x + 8)$$
$$x + 32 = 2x + 16$$
$$16 = x$$

$X + 8 \qquad X + 24$

Frank's present age is 16, and Eva's present age is 40.

Investment Problems

Investment problems usually involve interest and require more than simply calculating interest earned on a given principal amount at a given rate. They generally call for you to set up and solve an algebraic equation.

Example: Investment Problem

Q: Dr. Kramer plans to invest $20,000 in an account paying 6% interest annually. How many additional dollars must she invest at the same time at 3% so that her total annual income during the first year is 4% of her entire investment?

Letting x equal the amount invested at 3%, Dr. Kramer's total investment can be expressed as $20,000 + x$. The interest on $20,000 plus the interest on the additional investment equals the total interest from both investments. Stated algebraically:

$$.06(20,000) + .03x = .04(20,000 + x)$$

Multiply all terms by 100 to eliminate decimals, and solve for x:

$$6(20,000) + 3x = 4(20,000 + x)$$
$$120,000 + 3x = 80,000 + 4x$$
$$40,000 = x$$

She must invest $40,000 at 3% for her total annual income to be 4% of her total investment ($60,000)

In solving investment problems, remember:

- It's best to simply eliminate percentage signs (or multiply terms by 100 to eliminate decimals).

- Don't try to solve these problems intuitively; they can be misleading on their face. (For instance, in the example above, you may have guessed or "intuited" that Dr. Kramer would have to invest more than *twice* as much at 3% than at 6% to lower the overall interest rate to 4%. Not true!)

Quiz Time

Here are 10 problems to test your skill in applying the concepts in this chapter. After attempting all 10 problems, read the explanations that follow. Then go back to the chapter and review your trouble spots. If you can handle the easier problems *and* the more challenging ones, consider yourself a "smart test-taker"!

Easier

1. Of 60 employees at Microfirm Company, x employees are computer programmers. If $\frac{2}{3}$ of the remaining employees are salespeople, express the number of Microfirm employees who are neither programmers nor salespeople, in terms of x. Express your answer in simplest terms.

2. If a portion of $10,000 is invested at 6% and the remaining portion is invested at 5%, and if x represents the amount invested at 6%, express the annual income in dollars from the 5% investment in terms of x.

3. Jill is now 20 years old and her brother Gary is now 14 years old. How many years ago was Jill three times as old as Gary was at that time?

4. Lisa has 45 coins which are worth a total of $3.50. If the coins are all nickels and dimes, how many more dimes than nickels does she have?

5. Two buses are 515 miles apart. At 9:30 a.m. they start traveling toward each other at rates of 48 and 55 miles per hour. At what time will they pass each other?

More Challenging

6. An investor wishes to sell some of the stock that he owns in MicroTron and Dynaco Corporations. He can sell MicroTron stock for $36 per share, and he can sell Dynaco stock for $52 per share. If he sells 300 shares altogether at an average price per share of $40, how many shares of Dynaco stock has he sold?

7. Dan drove home from college at an average rate of 60 miles per hour. On his trip back to college, his rate was 10 miles per hour slower and the trip took him one hour longer than the drive home. How far is Dan's home from the college?

8. How many ounces of soy sauce must be added to 18 ounces of a peanut sauce and soy sauce mixture consisting of 32% peanut sauce to create a mixture which is 12% peanut sauce?

9. If a train travels $r + 2$ miles in h hours. Express the number of miles the train travels in 1 hour and 30 minutes in terms of r and h?

10. Two water hoses feed a 40-gallon tank. If one of the hoses dispenses water at the rate of 2 gallons per minute, and the other hose dispenses water at the rate of 5 gallons per minute, how many minutes does it take to fill the 40-gallon tank, if the tank is empty initially?

Answers and Explanations

1. The answer is $20 - \frac{1}{3}x$.

The number of non-programmers can be expressed as $60 - x$.

1/3 of these non-programmers are salespeople:

$$\frac{1}{3}(60 - x) = 20 - \frac{1}{3}x.$$

2. The answer is $500 - .05x$. The amount invested at 5% is $10,000 - x$ dollars. Thus, the income from that amount is $.05(10,000 - x)$ dollars.

3. Jill's age x years ago can be stated algebraically as $20 - x$. At that time, Gary's age was $14 - x$. The following equation emerges:

$$20x = 3(14 - x)$$
$$20 - x = 42 - 3x$$
$$2x = 22$$
$$x = 11$$

Jill was 3 times as old as Gary 11 years ago. (Jill was 9 and Gary was 3.)

4. The answer is 5. Letting x equal the number of nickels:

$$45 - x = \text{ the number of dimes}$$
$$5x = \text{ the value of all nickels (in cents)}$$
$$450 - 10x = \text{ the value of all dimes (in cents)}$$

Given a total value of 350 cents:

$$5x + 450 - 10x = 350$$
$$-5x = -100$$
$$x = 20$$

Lisa has 20 nickels and 25 dimes; thus, she has 5 more dimes than nickels.

5. The answer is 2:30 p.m. The total distance is equal to the distance traveled by one bus plus the distance traveled by the other bus (to the point where they pass each other). Letting x equal the number of hours traveled, the distances that the two buses travel in that time can be expressed as $48x$ and $55x$. Equating the sum of these distances with the total distance and solving for x:

$$48x + 55x = 515$$
$$103x = 515$$
$$x = 5$$

The buses will pass each other 5 hours after 9:30 a.m.—at 2:30 p.m.

6. The answer is 75. The value of Dynaco shares sold plus the value of MicroTron shares sold must be equal to the value of all shares sold (i.e., the "mixture"). Letting x represent the number of shares of Dynaco sold, the number of shares of MicroTron sold can be represented by $300 - x$. The following table represents all values algebraically:

	# of shares	price per share	total value
Dynaco	x	52	$52x$
MicroTron	$300 - x$	36	$36(300 - x)$
mixture	300	40	1,200

Set up an equation in which the value of Dynaco shares sold plus the value of MicroTron shares sold equals the total value of all shares sold, and solve for x:

$$\$52(x) + \$36(300 - x) = \$40(300)$$
$$52x + 10{,}800 - 36x = 12{,}000$$
$$16x = 1{,}200$$
$$x = 75$$

The investor has sold 75 shares of Dynaco stock. Checking your work:

$$\$52(75) + \$36(300 - 75) = \$12{,}000$$
$$\$3{,}900 + \$36(225) = \$12{,}000$$
$$\$3{,}900 + 8{,}100 = \$12{,}000$$

7. The answer is 300. The distance can be expressed both in terms of Dan's driving time going home and going back to college. Letting x equal the time (in hours) it took Dan to drive home, the distance between his home and his college can be expressed both as $60x$ and as $50(x + 1)$. Equating the two distances (since distance is constant) and solving for x:

$$60x = 50(x + 1)$$
$$60x = 50x + 50$$
$$10x = 50$$
$$x = 5$$

It took Dan 5 hours at 60 miles per hour to drive from college to home, thus, the distance is 300 miles.

8. The answer is 30. Letting x equal the number of ounces of soy sauce added to the mixture, $18 + x$ equals the total amount of the mixture after the soy sauce is added. All values can be represented algebraically in the following table:

	# of ounces	% of peanut sauce	amount of peanut sauce
original	18	32	5.76
added	x	0	0
new	$x + 18$	12	$12x + 216$

The amount of peanut sauce (5.76 ounces) must equal 12% of the new total amount of the mixture, which is $18 + x$. Expressed as an algebraic equation and solving for x:

$5.76 = .12(x + 18)$

$576 = 12(x + 18)$

$576 = 12x + 216$

$360 = 12x$

$x = 30$

30 ounces of soy sauce must be added to result in a mixture that is 12% peanut sauce.

9. The answer is $\frac{3r + 6}{2h}$. Given that the train travels $r + 2$ miles in h hours, its rate in miles per hour can be expressed as $\frac{r + 2}{h}$. In $1\frac{1}{2}$ hours, the train would travel $\frac{3}{2}$ this distance, or $\left(\frac{3}{2}\right)\left(\frac{r + 2}{h}\right) = \frac{3(r + 2)}{2h} = \frac{3r + 6}{2h}$.

10. The answer is $5\frac{5}{7}$ minutes. The first hose can perform $\frac{1}{20}$ of the job in one minute. The second hose can perform $\frac{1}{8}$ of the job in one minute. The two rates can be added together to obtain the aggregate rate per minute:

$$\frac{1}{20} + \frac{1}{8} = \frac{1}{A}$$

$$\frac{2}{40} + \frac{5}{40} = \frac{1}{A}$$

$$\frac{7}{40} = \frac{1}{A}$$

$$A = 5\frac{5}{7}$$

It takes $5\frac{5}{7}$ minutes for both hoses together to fill the tank.

Here's an alternative explanation: Together, the hoses fill the tank at a combined rate of 7 gallons per minute. Letting x equal the total time: $7x = 40$, or $x = 5\frac{5}{7}$.

13

Geometry
—Lines, Angles, and Triangles

After ten chapters of crunching numbers and bending and twisting equations, you're probably in pretty good numeric and algebraic shape. Now it's time to get into *shapes*—geometric shapes, that is. On average, geometry problems account for about 20% of all math-type problems on standardized exams. Here's a quick look at the topics covered on the exams:

2-dimensional shapes (on a plane), including:

- intersecting lines and parallel lines
- triangles
- 4-sided shapes or "quadrilaterals" (squares, rectangles, parallelograms, trapezoids)
- circles

3-dimensional shapes, or "solids," including:

- rectangular solids (cubes and other "blocks")
- cylinders
- pyramids

Coordinate geometry (using the Cartesian grid or plane)

A typical geometry problem might require you to determine:

- the size (or measure) of an angle—in degrees

- the length of one side of an object

- the distance around an object—that is, its perimeter (or for a circle, circumference)

- the area of a two-dimensional figure (either part of it or all of it)

- the volume of a three-dimensional figure (either part of it or all of it)

Some problems will go further, requiring you to compare angle measures, lengths, areas, or volumes, or to recognize the relationship between an angle measure and a length, a length and an area, and so forth.

For all of these tasks, you're going to need to know, here's that awful word again: *formulas.* As with algebra word problems, though, the good news is that you don't need to know very many. In the next three chapters, you'll examine each topic listed above (in that order), learning the formulas you need to handle the exam problems with no sweat. Let's start with geometry problems involving intersecting lines and triangles.

Lines and Angles

Lines and *line segments* are the basic building blocks for all geometry problems. In fact, some geometry problems involve nothing more than intersecting lines (and the angles created thereby). Two types are most common on the exams:

- problems involving **wheel-spokes**

- problems involving **parallel lines** and **transversals**

Before examining these two problems types, let's review some basic terminology and some fundamental rules. Referring to the diagram on page 219, the following symbols are used on the exams to denote lines, line segments, and angles that result from intersecting lines:

- l_1 identifies one line, and l_2 identifies the other. (Lines and line segments are always assumed to be straight.)

- *AE* identifies line segment *AE*; it can also be used to denote the *length* of line segment *AE*.

- point *E* is the *vertex* of any of the angles formed by the intersection of l_1 and l_2

- ∠*AEB* denotes the angle having point *E* as its vertex and *EA* and *EB* as the two line segments forming the angle. The size of ∠*AED* is *x*°—that is, ∠*AED* has a degree measure of *x*.

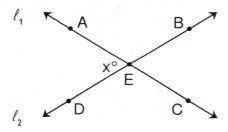

Opposite angles are the same size, or *congruent* (≅), or equal in degree measure. In the diagram above:

$$\angle AED \cong \angle BEC$$

$$\angle AEB \cong \angle DEC$$

Supplementary angles form a straight line when added together; Their degree measures total 180. In fact, a straight line is actually a 180° angle. In the diagram above:

$$\angle AED + \angle AEB = 180° \quad \text{a straight line}$$

$$\angle AEB + \angle BEC = 180° \quad \text{a straight line}$$

$$\angle BEC + \angle CED = 180° \quad \text{a straight line}$$

$$\angle CED + \angle AED = 180° \quad \text{a straight line}$$

Wheel Spokes

Building on the rules you just learned, the sum of all angles forming a circle is 360° regardless of how many angles are involved. Thus, in the diagram above:

$$\angle AEB + \angle BEC + \angle CED + \angle AED = 360°$$

A *right angle* is an angle measuring 90°. The intersection of two *perpendicular* lines results by definition in a right angle. If you know that one angle formed by two intersecting lines is a right (90°) angle, then you know that the two lines are perpendicular and that all four angles formed by the intersection are right angles, as illustrated below:

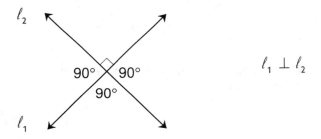

Combining all of the foregoing rules, the following relationships among the angles in the diagram below emerge (the last two are less obvious and therefore more "testworthy"):

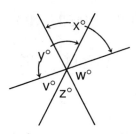

$y + v = 180$	y and v form a straight line—they are supplementary
$x + w = 180$	x and w form a straight line—they are supplementary
$v + z + w = 180$	v, z and w form a straight line—they are supplementary
$x + y - z = 180$	$x + y$ exceeds 180 by the amount of their overlap, which equals z—the angle opposite to the overlapping angle
$x + y + v + w = 360$	the sum of all angles, excluding z, is $360°$; z is excluded because it is already accounted for by the overlap of x and y

Parallel Lines and Transversals

Parallel lines are, of course, lines that never intersect, even continuing infinitely in both directions. On a math exam, two parallel lines, 1 and 2, would be indicated this way:

$$l_1 \parallel l_2$$

Exam problems involving parallel lines also involve at least one *transversal*, which is a line that intersects each of two (or more) parallel lines. In the figure on page 222, $l_1 \parallel l_2$, and l_3 *transverses* (is a transversal of) l_1 and l_2:

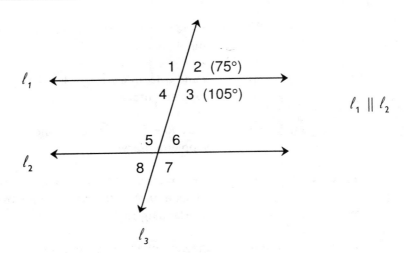

In the figure above, because $l_1 \| l_2$, the upper "cluster" of angles (created by the intersection of l_1 and l_3) looks identical to, or "mirrors," the lower "cluster" (created by the intersection of l_2 and l_3). For example, $\angle 1$ is congruent (equal in size or degree measure) to $\angle 5$ ($\angle 1$ and $\angle 5$ are said to be *corresponding* angles). Because opposite angles are congruent:

- all of the *odd*-numbered angles are congruent (equal in size) to one another

- all of the *even*-numbered angles are congruent (equal in size) to one another

Moreover, if you know the size of just one of the eight angles, you can determine the size of all eight angles. For example, if $\angle 2$ measures 75°, then angles 4, 6 and 8 also measure 75° each, while angles 1, 3, 5 and 7 each measure 105° (75° + 105° = 180°, a straight line).

If a second transversal paralleling the first one is added, the resulting four-sided figure is a *parallelogram*—a quadrilateral with two pairs of parallel sides. Applying the transversal analysis to parallelogram *ABCD* on page 223, it is evident that:

$$\angle 1 \cong \angle 4$$

$$\angle 2 \cong \angle 3$$

$$\angle 1 + \angle 2 = 180°,\ \angle 1 + \angle 3 = 180°$$
$$\angle 2 + \angle 4 = 180°,\ \angle 3 + \angle 4 = 180°$$

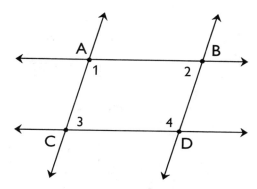

In fact, all four "clusters" of angles (defined by vertices A,B,C and D) mirror one another in their corresponding angle measures. If you know the size of just one angle, you can determine the size of all 16 angles!

Triangles

The *triangle* (defined as a three-sided polygon) is the test-makers' favorite geometric figure. All of the geometry topics examined in this chapter and in the next two are covered on the exams and therefore important, but if you must prioritize your study, pay special attention to triangles. Why? Because you often need to understand triangles to solve other geometry problems, including some involving 4-sided figures, three-dimensional solids, and even circles. **Remember:** If you don't understand those 3-sided shapes, expect to have trouble with others as well.

In this section, after examining the properties that apply to all triangles, we'll focus on three special types of triangles—right, isosceles, and equilateral—which are the most testworthy of all. If you're ready, let's triangulate!

Properties of All Triangles

Let's begin by looking at the following two triangles:

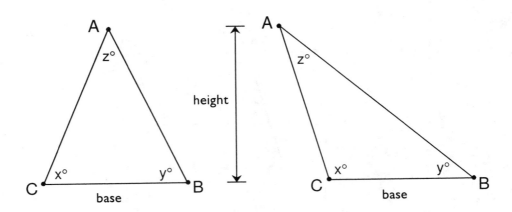

Notice that their shapes are quite different from each other—both in terms of angle sizes and side lengths. They're drawn differently to help make clear that the five basic characteristics—or properties listed below—apply to all triangles, regardless of shape or size.

Length of sides. This first characteristic is perhaps the most obvious one. Each side of a triangle is smaller in length than the sum of the lengths of the other two sides; for example:

$$\overline{AC} < \overline{AB} + \overline{BC}$$

$$\overline{BC} < \overline{AB} + \overline{AC}$$

$$\overline{AB} < \overline{BC} + \overline{AC}$$

This makes good sense if you stop to think about it. What if two sides together were equal to a third side in length? The triangle would collapse, wouldn't it? You'd have two overlapping segments.

Angle measures. The sum of the three angles of any triangle $= 180°$

$$x + y + z = 180$$

It follows, then, that 180 minus the measure of one angle must equal the sum of the other two angles. Why? Just subtract either x, y or z from the equation above:

$$180 - x = y + z$$
$$180 - y = x + z$$
$$180 - z = x + y$$

Since x, y, and z each must be greater than zero (we couldn't form a triangle otherwise), the sum of the measures of any two (of the three) angles must be less than $180°$:

$$x + y < 180 \ (x < 180 - y, \ y < 180 - x)$$
$$x + z < 180 \ (x < 180 - z, \ z < 180 - x)$$
$$y + z < 180 \ (y < 180 - z, \ z < 180 - y)$$

Angles and opposite sides. In any triangle, the relative angle sizes correspond to the relative lengths of the sides opposite those angles. In other words, the smaller the angle, the smaller the side opposite the angle, and vice versa. Accordingly, if two angles are equal in size, the sides opposite those angles are of equal length (and vice-versa). Referring to the lefthand triangle on page 224, let's assume that $AC = AB$ (it looks that way, anyway). If so, then $x = y$. In the righthand triangle, let's assume that AB is the longest side, CB is the shortest side, and AC is in between the other two sides in length (it looks that way, anyway). Accordingly, $x > y > z$. To help you visualize this, in the diagram at the top of page 226 some numbers have been assigned to the sides of our two triangles.

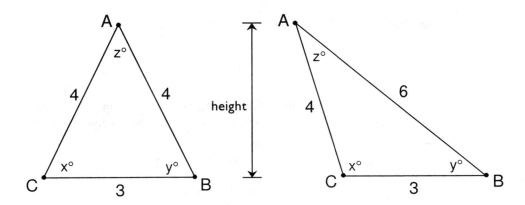

Caution: Do not take this rule too far. The relative size of angle measures does *not* correspond linearly to the relative size of opposite sides! In the two triangles above, although the relative lengths of the three sides are 4:4:3 and 3:4:6, the angles opposite these sides are not in the proportionate size to one another. Here's another example: if a certain triangle has angle measures of 30°, 60°, and 90°, the ratio of the angles is 1:2:3. However, this does *not* mean that the ratio of the opposite sides is also 1:2:3 (it is NOT, as you will soon learn).

Area. The area of any triangle is equal to $\frac{1}{2}$ the product of its base and its height, also called the *altitude*:

$$A_t = \frac{1}{2}(b)(h)$$

Any side can be used as the base to calculate area. In the two triangles at the top of page 227, notice that the "bottom" side is labeled as the "base." However, you can just as easily push either triangle over so that another side becomes the base. Also notice the height is not the length of any particular side; height is the triangle's altitude—how far up from the base it extends. Imagine the base on flat ground; drop a plumb line straight down from the top peak of the triangle to define height or altitude:

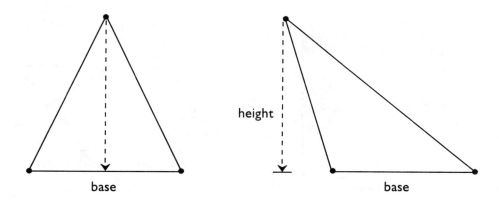

The only case where the height of a triangle will equal the length of any of its sides is with a triangle having a 90° angle (called a right *triangle*):

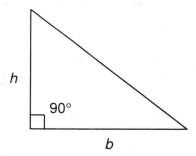

Here's some insight about the formula for a triangle's area $[\frac{1}{2}(base \times height)]$. If we ignore the $\frac{1}{2}$ and compute base × height, we would be determining the area of a rectangle with twice the area of the triangle. (At the risk of getting ahead of ourselves here, the area of a rectangle equals its base multiplied by its height—or length times width.) We can easily visualize this with the first of our two original triangles, although we need to turn the second one onto a different base to visualize a rectangle.

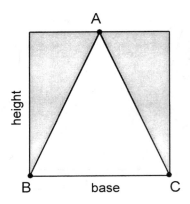

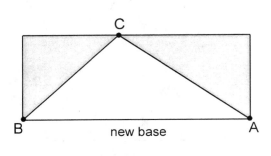

So, if you can create a rectangular box from a triangle of any shape—just high enough and wide enough to accommodate the triangle, the box will have twice the area of the triangle!

To appreciate this point, pretend the triangle is made of string, and slide $\angle C$ to the left and right along the top of the box:

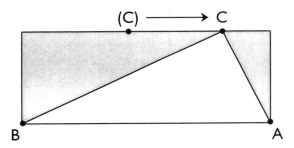

Sliding $\angle C$ horizontally does not change the area of the triangle or the fact that its area is half that of the rectangular container!

Right Triangles and the Pythagorean Theorem

You are about to enter the inner sanctum of geometry—where that very favorite of the test makers' arsenal of shapes resides. I'm referring to the Right Triangle (capitalized here just for dramatic impact).

In a *right triangle*, one angle measures 90° and, of course, each of the other two angles measures less than 90°. The two sides forming the 90° angle are commonly referred to as the triangle's *legs* (a and b below), while the third (and longest side) is referred to as the *hypotenuse* (c below).

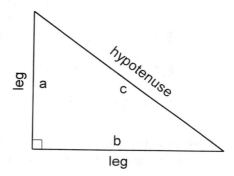

The relative lengths of the sides of any right triangle can be expressed by the *Pythagorean Theorem* (a and b are the two legs, and c is the hypotenuse):

$$a^2 + b^2 = c^2 \text{ or } \sqrt{a} + \sqrt{b} = \sqrt{c}$$

The Pythagorean Theorem is the single most useful formula for the test makers devising geometry problems! With any right triangle, if you know the length of two sides, you can determine the length of the third side with the Theorem. *Remember:* The Pythagorean Theorem applies only to *right* triangles, not to any others!

Pythagorean Side Triplets

These are sets of numbers that satisfy the Pythagorean Theorem. In each triplet in the chart at the top of page 230, the first two numbers represent the relative lengths of the two legs, while the third—and largest—number represents the relative length of the hypotenuse:

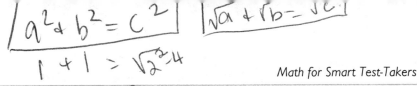

$$a^2 + b^2 = c^2 \qquad \sqrt{a} + \sqrt{b} = \sqrt{c}$$

$$1 + 1 = \sqrt{2}\cdot4$$

$$1 + 1 = 2$$

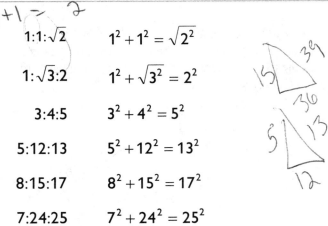

$1:1:\sqrt{2}$	$1^2 + 1^2 = \sqrt{2}^2$
$1:\sqrt{3}:2$	$1^2 + \sqrt{3}^2 = 2^2$
$3:4:5$	$3^2 + 4^2 = 5^2$
$5:12:13$	$5^2 + 12^2 = 13^2$
$8:15:17$	$8^2 + 15^2 = 17^2$
$7:24:25$	$7^2 + 24^2 = 25^2$

Each triplet above is expressed as a *ratio* since it represents the relative proportion of the triangle's sides. All right triangles with sides having the same ratio or proportion have the same shape (they are *similar* to one another). For example, a right triangle with sides 5, 12, and 13 units in length is smaller but exactly the same shape (proportion) as one with sides 15, 36 and 39 units in length. Learn to recognize given numbers (lengths of triangle sides) as multiples of Pythagorean triplets to save valuable time in solving right-triangle problems. Let's look at some examples.

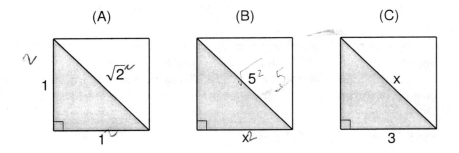

(A)   (B)   (C)

All three squares above include two $1:1:\sqrt{2}$ triangles.

(A) indicates the basic triplet.

(B) The hypotenuse is given as 5. To calculate the length of either leg, divide 5 by $\sqrt{2}$, *or* multiply 5 by $\dfrac{\sqrt{2}}{2}$ $\left(x = 5 \times \dfrac{\sqrt{2}}{2} = 5\dfrac{\sqrt{2}}{2}\right)$.

$\sqrt{27} + 3^2 = x^2$

(C) The length of leg is given as 3. To calculate the hypotenuse, multiply 3 by $\sqrt{2}$ $(3 \times \sqrt{2} = x = 3\sqrt{2})$.

(A)   (B)   (C)

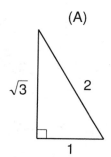
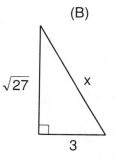
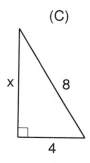

All three triangles above are $1:\sqrt{3}:2$ triangles.

(A) indicates the basic triplet.

(B) The length of the legs are given. $\sqrt{27} = 3\sqrt{3}$, and the ratio of $3\sqrt{3}$ to 3 is $\sqrt{3}:1$. Thus, this is a $1:\sqrt{3}:2$ triangle. To calculate the hypotenuse, either:

multiply 3 by 2 $(x = 3 \times 2 = 6)$

or multiply $\sqrt{27}$ by $\dfrac{2}{\sqrt{3}}$ $\left(3\sqrt{3} \times \dfrac{2}{\sqrt{3}} = 6\right)$.

(C) The length of the hypotenuse and of one leg are given as 8 and 4. This ratio is $2:1$. Thus, this is a $1:\sqrt{3}:2$ triangle. To calculate the length of leg x, either:

multiply 4 by $\sqrt{3}$ $(x = 4\sqrt{3})$

or multiply 8 by $\dfrac{3}{\sqrt{2}}$ $\left(x = 4\sqrt{3}\right)$.

225

1296

$a^2 + b^2 = c^2$

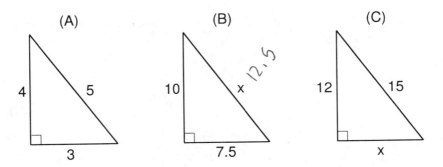

All three triangles above are 3:4:5 triangles.

(A) indicates the basic triplet.

(B) The length of the legs are given. The ratio of 10 to 7.5 is 4:3. Thus, this is a 3:4:5 triangle. To calculate the hypotenuse, either:

multiply 7.5 by $\frac{5}{3}$ ($x = 12.5$)

or multiply 10 by $\frac{5}{4}$ ($x = 12.5$)

(C) The length of the hypotenuse and of one leg are given as 15 and 12. This ratio is 5:4. Thus, this is a 3:4:5 triangle. To calculate the length of leg x, either:

multiply 15 by $\frac{3}{5}$ ($x = 9$)

or multiply 12 by $\frac{3}{4}$ ($x = 9$)

Pythagorean Angle Triplets

In two (and only two) of the unique triangles identified above as triplets, *all degree measures are integers*:

- The corresponding angles opposite the sides of a $1:1:\sqrt{2}$ triangle are 45°, 45° and 90°

- The corresponding angles opposite the sides of a $1:\sqrt{3}:2$ triangle are 30°, 60° and 90°

These angle triplets and their corresponding leg triplets are illustrated below:

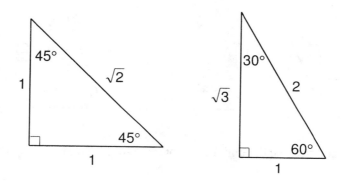

Thus, to quickly determine *all* angle measures and lengths of *all* sides of a triangles, all you need to know is:

1. that the triangle is a right triangle;

2. that one of the angles is either 30°, 45° or 60°; and

3. the length of any one of the three sides.

Isosceles Triangles

An *isosceles* triangle is one in which at least two sides are equal in length—and, accordingly, at least two angles are equal in size. In any isosceles triangle, an *altitude* line from the angle formed by the equal sides always bisects the opposite side. Thus, if you know the lengths of the sides, you can easily determine the triangle's area by applying the Pythagorean Theorem to determine the height. Consider, for example, the Isosceles triangle *(A)* at the top of page 234:

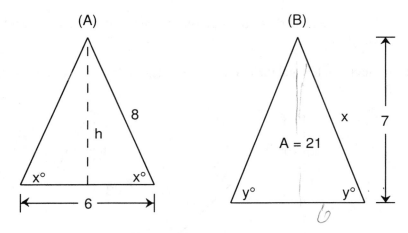

The height can be determined by applying the Pythagorean Theorem:

$$3^2 + h^2 = 8^2$$

$$h^2 = 64 - 9 = 55$$

$$h = \sqrt{55}$$

Thus, the area of the triangle $A = \frac{1}{2}(6)(\sqrt{55}) = 3\sqrt{55}$. If you knew h but did not know that the length of two of the sides were each 8 (as indicated above), you could determine that length. The altitude line h bisects the base AB, creating two congruent right triangles, each with legs of length 3 and $\sqrt{55}$. Applying the Pythagorean Theorem:

$$x^2 = 3^2 + \left(\sqrt{55}\right)^2$$

$$x^2 = 9 + 55 = 64$$

$$x = 8$$

In isosceles triangle B above, the area of the triangle is given as 21, and its height is 7. You can determine x by applying the area formula for a triangle, then the Pythagorean Theorem:

$$A = \frac{1}{2}(b)(h) \qquad\qquad x^2 = 3^2 + 7^2$$

$$21 = \frac{1}{2}(b)(7) \qquad\qquad x^2 = 9 + 49 = 58$$

$$b = 6 \qquad\qquad\qquad x = \sqrt{58}$$

Equilateral Triangles

An *equilateral* triangle is a special type of isosceles triangle in which all three sides are the same length—and, accordingly, all three angles are the same size (60°). Any line bisecting one of the 60° angles will divide an equilateral triangle into two right triangles with angle measures of 30, 60, and 90°—i.e., two $1:\sqrt{3}:2$ triangles.

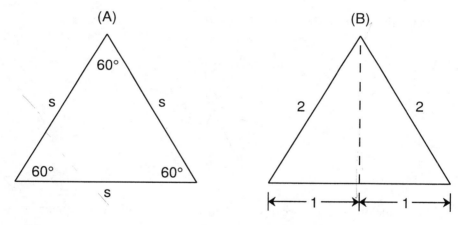

The area of an equilateral triangle $= \left(\dfrac{s^2}{4}\right)\left(\sqrt{3}\right)$, where s is the length of a side. In the

figure above, if $s = 6$, the area of the triangle $= 9\sqrt{3}$. To confirm this formula, bisect the triangle into two 30-60-90° ($1:\sqrt{3}:2$) triangles (as in diagram *b* above).

The area of the equilateral triangle is $\frac{1}{2}(2)(\sqrt{3})$ or $\sqrt{3}$. [The area of each smaller right

triangle is $\dfrac{\sqrt{3}}{2}$ (remember the $1:\sqrt{3}:2$ triplet?).]

Note: On the exams, you are most likely to encounter equilateral triangles in problems involving *circles* (more on this in Chapter 12).

Quiz Time

Here are 10 problems to test your skill in applying the concepts in this chapter. After attempting all 10 problems, read the explanations that follow. Then go back to the chapter and review your trouble spots. If you can handle the easier problems *and* the more challenging ones, consider yourself a "smart test-taker"!

Easier

1.   In the figure below, if $y = 100$ and $z = 135$, what is the value of x?

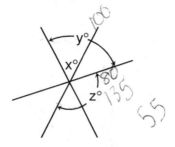

2.   In parallelogram *ABCD* below, $\angle A$ measures 60°. What is the sum of $\angle B$ and $\angle D$?

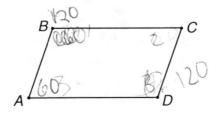

3. Each number pair below indicates the length of the hypotenuse and one of the other two sides of a right triangle. For each pair, determine the length of the remaining side.

(a) $3.5\sqrt{3}$ and 7

(b) 2 and $\sqrt{2}$

(c) 20 and 16

(d) 26 and 10

4. A rectangular door measures 5 feet by 6 feet 8 inches. What is the distance from one corner of the door to the diagonally-opposite corner?

5. Assume you are given a yardstick (36 inches long) and told to saw it into three pieces to form a triangular enclosure. What is the largest possible area of such an enclosure?

More Challenging

6. In the figure below, what is the value of x ?

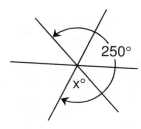

7. In $\triangle ABC$, $AB = BC$. If the size of $\angle B$ is x degrees, express the degree measure of $\angle A$ in terms of x?

8. In $\triangle RST$ below, If A is the midpoint of RS and B is the midpoint of ST, then which of the following must be true?

(A) $SA > ST$ (B) $BT > BS$ (C) $BT = SA$ (D) $SR = RT$ (E) $RT > ST$

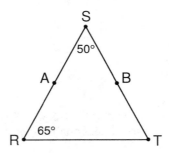

9. In the figure below, if $BE = EC$, what is the area of $\triangle CDE$?

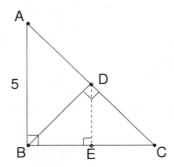

10. What is the perimeter of an equilateral triangle whose area is $16\sqrt{3}$?

Answers and Explanations

1. $x = 55$. Angles y and z exceed 180° by the value of x; that is, $y + z - x = 180°$. Substitute 100 and 135 for y and z, respectively, in order to solve for x.

2. The answer is 240°. Given that $\angle A = 60°$, $\angle B$ must equal 120°, since the angle measures must total 180°. $\angle B = \angle D$. Thus, their sum is 240°.

3. Each pair includes two of three numbers in a Pythagorean triplet:

   (a) $3.5:3.5\sqrt{3}:7$ (basic triplet is $1:\sqrt{3}:2$)

   (b) $\sqrt{2}:\sqrt{2}:2$ (basic triplet is $1:1:\sqrt{2}$)

   (c) $12:16:20$ (basic triplet is $3:4:5$)

   (d) $10:24:26$ (basic triplet is $5:12:13$)

4. The answer is 8 feet 4 inches. The width of the door is 60 inches (5 feet), and its length is 80 inches (6 feet 8 inches). This is a 6:8:10 triangle (conforming to the 3:4:5 Pythagorean triplet), with a diagonal of 100 inches, or 8 feet 4 inches.

5. The answer is $36\sqrt{3}$. This question calls for some intuition. You should make your two cuts at the 12-inch and 24-inch marks. All three pieces should be equal in length (12 inches each) to form an equilateral triangle. Any other shape would result in a "flatter" triangle with a smaller-than-maximum area. Since each side is 12 inches in length, the area is $36\sqrt{3}$.

6. $x = 70$. Referring to the figure below, the total degree measure of all angles is 360. Given that all angles but a and b total 250, $a + b = 110$. $a + b + x = 180$ (they form a straight line). Thus, $x = 70$.

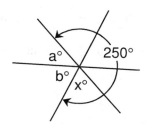

7. The answer is $90 - \frac{x}{2}$. The triangle is isosceles, so $\angle A = \angle C$. Letting a, c and x represent the degree measures of $\angle A$, $\angle C$ and $\angle B$, respectively, solve for a:

$$a + c + x = 180$$
$$2a + x = 180 \; [a = c]$$
$$2a = 180 - x$$
$$a = \frac{180}{2} - \frac{x}{2}$$
$$a = 90 - \frac{x}{2}$$

8. The correct answer is (C). Given that $\angle R + \angle S = 115°$ and that the sum of the three angles equals 180, $\angle T$ must equal $65°$. The triangle is isosceles, and $RS = ST$. Given that $BT = \frac{1}{2} \, ST$ and that $SA = \frac{1}{2} \, SR$, BT must equal SA.

9. The answer is 3.125 (or $3\frac{1}{8}$). Without resorting to a formal proof, this question calls for a bit of intuition and visualization. Because $BE = EC$, D must bisect AC, and $\triangle ABC$ is isosceles. In fact, all of the triangles in the diagram (there are actually five of them) are right isosceles triangles. The area of the largest triangle ($\triangle ABC$) is exactly twice the area of either $\triangle ABD$ or $\triangle BCD$. The area of each of these two triangles is exactly twice the area of $\triangle BDE$ and $\triangle CDE$. Thus, the area of $\triangle CDE$ is one-fourth the area of the largest triangle ($\triangle ABC$). The area of $\triangle ABC$ is 12.5. Thus, the area of $\triangle CDE = 3.125$.

10. The answer is 24. The area of an equilateral triangle $= \frac{s^2}{4}\sqrt{3}$. Therefore, $\frac{s^2}{4} = 16$. $s^2 = 64$, and $s = 8$. The perimeter is $8 + 8 + 8 = 24$.

14

Geometry
—Quadrilaterals and Circles

In this chapter we'll "square off" against four-sided polygons—known as *quadrilaterals*, as well as round out our study of two-dimensional figures by examining other polygons and the ubiquitous *circle*. The specific types of quadrilaterals that appear most frequently on math exams are:

the **square**, which is a special type of…

rectangle, which is a special type of…

parallelogram, which is a special type of **quadrilateral**.

Although the following two types of quadrilaterals appear less frequently on the exams, you should also be familiar with them:

the **rhombus**

the **trapezoid**

Each of these quadrilaterals, as well as the circle, has its own properties, or characteristics that should be second nature to you as you approach your exam. The two most important properties are:

area (surface covered by the figure on a plane) and

perimeter (total length of all sides) or, in the case of circles, the *circumference* (the total distance around the circle)

We'll get into circles a bit later. For now, let's focus on "quads." The figure below indicates the perimeter and area formulas for each of the five quadrilaterals listed above. ("A" indicates area, and "P" indicates perimeter.) Memorize these formulas!

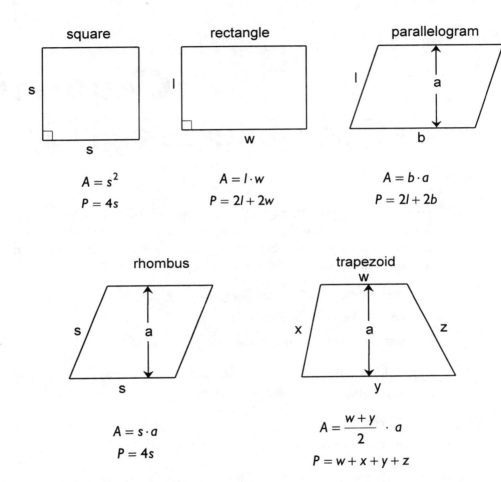

Characteristics of a Square

Unless a quadrilateral has all 4 of these characteristics, it ain't square (#2 is the only one that applies to any rectangle):

- all 4 sides are equal in length

- all 4 angles are right angles (90°)

- perimeter $= 4s$

- area $= s^2$

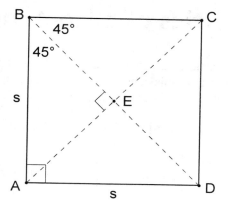

The dotted lines in square *ABCD* above are called *diagonals*, a term meaning line segments that connect opposite corners of a quadrilateral. When diagonals are added to a square:

- the area of square $= \dfrac{(AC)^2}{2}$ or $\dfrac{(BD)^2}{2}$ (diagonal squared, divided by 2—this formula applies only to squares, not to other quadrilaterals!)

- the diagonals are equal in length ($\overline{AC} = \overline{BD}$)

- the diagonals are perpendicular; their intersection creates 4 right angles

- the diagonals *bisect* each 90° angle of the square; that is, they split each angle into two equal (45°) angles

- 4 distinct *congruent* (the same shape and size) triangles, each having an area of $\dfrac{1}{2}$ that of the square, are created: $\triangle ABD$, $\triangle ACD$, $\triangle ABC$, $\triangle BCD$

- 4 distinct congruent triangles, each having an area of $\frac{1}{4}$ that of the square, are created: $\triangle ABE$, $\triangle BCE$, $\triangle CDE$, $\triangle ADE$

- all 8 triangles created are *right isosceles* triangles (one angle measures 90° and two sides are equal in length).

Characteristics of a Rectangle

The key characteristic of a rectangle is that all four angles are 90° in size (they are right angles). As long as this condition is satisfied, then you've got a rectangle, and all of the following characteristics apply as well:

- opposite sides are equal in length

- perimeter $= 2l + 2w$

- area $= l \times w$

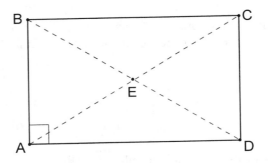

The maximum area of a rectangle with a given perimeter is a square. Conversely, the minimum perimeter of a rectangle with a given area is a square.

If you add diagonals to a rectangle (as in $ABCD$ above):

- the diagonals are equal in length ($AC = BD$)

- the diagonals are *not* perpendicular (unless the rectangle is a square)

- the diagonals do *not* bisect each 90° angle of the rectangle (unless the rectangle is a square)

- $AE = BE = CE = DE$

- 4 distinct congruent triangles, each having an area of $\frac{1}{2}$ that of the rectangle, are created: $\triangle ABD$, $\triangle ACD$, $\triangle ABC$, $\triangle BCD$

- $\triangle ABE$ is congruent to $\triangle CDE$; both triangles are isosceles (but they are right triangles *only* if the rectangle is a square)

- $\triangle BEC$ is congruent to $\triangle AED$; both triangles are isosceles (but they are right triangles *only* if the rectangle is a square)

- 4 distinct congruent right triangles, each having an area of $\frac{1}{4}$ that of the square, are created: $\triangle ABE$, $\triangle BEC$, $\triangle CED$, $\triangle AED$

Characteristics of a Parallelogram

Parallelograms are just like rectangles, except that the angles need not be 90° (a rectangle is a parallelogram with 90° angles). Here are the characteristics of any parallelogram:

- opposites sides are parallel

- opposite sides are equal in length

- opposite angles are congruent (equal in degree measure)

- all 4 angles are congruent *only* if the parallelogram is a rectangle—i.e., if the angles are right angles

- perimeter $= 2l + 2w$

- area $=$ base $(b) \times$ altitude (a)

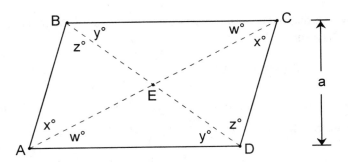

Adding diagonals to a parallelogram (as in *ABCD* above):

- the diagonals ($\overline{AC}$ and $\overline{BD}$) are *not* equal in length (unless the figure is a rectangle)

- the diagonals are *not* perpendicular (unless the figure is a square or rhombus)

- the diagonals do *not* bisect each angle of the parallelogram (unless it is a square or rhombus)

- the diagonals bisect each other ($\overline{AE} = \overline{ED}$, $\overline{CE} = \overline{AE}$)

- two pairs of congruent triangles are created, each having an area of $\frac{1}{2}$ that of the square: $\triangle ABD$ is congruent to $\triangle BCD$, $\triangle ACD$ is congruent to $\triangle ABC$

- $\triangle ABE$ is congruent to $\triangle CED$ (they are mirror-imaged horizontally *and* vertically); the triangles are isosceles only if the quadrilateral is a rectangle

- $\triangle BEC$ is congruent to $\triangle AED$ (they are mirror-imaged horizontally *and* vertically); the triangles are isosceles only if the quadrilateral is a rectangle

Characteristics of a Rhombus

One way to describe a rhombus is to say that it looks like a square that leans to one side. Here are the characteristics of a rhombus:

- all sides are equal in length

- opposite sides are parallel

- none of the four angles are right angles (angle measures $\neq 90°$)

- perimeter $= 4s$

- area $=$ side $(s) \times$ altitude (a)

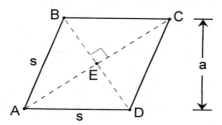

Adding diagonals (AC and BD) to a rhombus:

- area $= \dfrac{AC \times BD}{2}$ ($\dfrac{1}{2}$ the product of the diagonals—this formula applies to a rhombus and a square, but not to any other quadrilaterals!)

- the diagonals bisect each other ($BE = ED = AE = EC$)

- intersection of diagonals creates 4 right angles (diagonals are perpendicular)

- diagonals are *not* equal in length ($AC \neq BD$)

- diagonals bisect each angle of the rhombus

- two pairs of congruent (the same shape and size) isosceles triangles are created, each triangle having an area of $\dfrac{1}{2}$ that of the rhombus, ($\triangle ABD$ is congruent to $\triangle BCD$, $\triangle ACD$ is congruent to $\triangle ABC$; none of the 4 triangles are right triangles)

- $\triangle ABE$ is congruent to $\triangle CED$; both are right triangles (but not isosceles)

- $\triangle BEC$ is congruent to $\triangle AED$; both are right triangles (but not isosceles)

Characteristics of a Trapezoid

A trapezoid looks like a rectangle that has been pinched together at the top or bottom. Here are the characteristics:

- only one pair of opposite sides is parallel ($BC \parallel AD$)

- perimeter $= AB + BC + CD + AD$

- area $= \dfrac{BC + AD}{2} \times$ altitude (a), which is half the sum of the two parallel sides multiplied by the altitude

Trapezoids don't appear quite as often on exams as the other quadrilaterals examined above. They aren't as appealing to the test makers because they aren't as versatile.

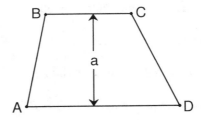

No predictable patterns emerge from the addition of two diagonals to a trapezoid.

Other Polygons

Among all the polygons in this polygonal world, we've already encountered just about all the varieties that appear on general standardized math exams. To ensure you've covered all the bases, though, let's look briefly at polygons having more than four sides. Consider the three polygons pictured below:

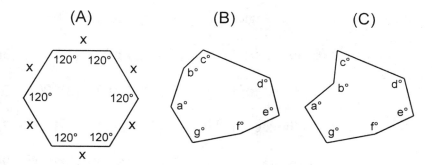

Notice in polygon (A), which is a hexagon, that when you "even up" the angles so that they're all the same, the lengths of the sides are also the same. Here's the rule, then: In any polygon, if the angles are equal in size, then all sides are equal in length, and you've got what's called a *regular* polygon. This rule makes sense when you think about it in the context of triangles, in which angles opposite equal sides are also equal in size. We're just expanding that rule to embrace all polygons. (It's a bit more complicated than that, but all you need to remember is the preceding rule.)

That's half of what you need to know about polygons with 5 or more sides. Here's the other half. Except for polygons such as polygon (C) above (where one of the interior angles exceeds 90°), you can use the following formula to determine the *sum* of all interior angles of a polygon (n = the number of side):

$$(n-2)180° = \text{sum of interior angles}$$

This applies to irregular polygons, such as septagon (B) above, as well as regular polygons such as hexagon (A). The chart on the next page applies this formula to polygons with sides numbering three through seven:

number of sides (polygon type)	sum of the interior angles
3 (triangle)	$(3 - 2)(180°) = 180°$
4 (quadrilateral)	$(4 - 2)(180°) = (2)180° = 360°$
5 (pentagon)	$(5 - 2)(180°) = (3)180° = 540°$
6 (hexagon)	$(6 - 2)(180°) = (4)180° = 720°$
7 (septagon)	$(7 - 2)(180°) = (5)180° = 900°$
8 (octagon)	$(8 - 2)(180°) = (6)180° = 1{,}080°$

Remember: This formula applies to both regular and irregular polygons. (It doesn't matter whether the sides or angles measures are equal.) However, it does *not* apply to any polygon with one or more *obtuse* interior angles (angles exceeding 90°).

Circles

You know what a circle looks like, of course. But how would you define it? As a two-dimensional "round thing"? Try this: A *circle* is the set of all points that lie equidistant from the same point (the circle's *center*) on a plane. Here are some other terms you should know:

radius: the distance from a circle's center to any point on the circle (*OA* and *OB* in the figure on page 251)

diameter: the greatest distance from one point to another on the circle (*AC* in the diagram on page 251); a circle's diameter is twice the length of its radius ($d = 2r$).

chord: a line segment connecting two points on the circle; the longest possible chord of a circle passes through its center and is the circle's diameter. (In the diagram on page 251, line segments *AB* and *AC* are both chords.)

circumference: the distance around the circle (its "perimeter")

arc: a segment of a circle's circumference (The figure on page 251 includes arc *AB* as well as chord *AB*.)

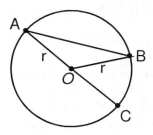

Determining a Circle's Area and Circumference

Here are the formulas for determining a circle's *circumference* and *area*:

$$\text{circumference} = 2\pi r \text{ or } \pi d$$

$$\text{area} = \pi r^2$$

With these two formulas, you can determine a circle's area, circumference, diameter, and radius, as long as you know just one of these four values. Here are two illustrations:

- A circle whose radius is 4 has a diameter of 8, a circumference of 8π, and an area of 16π

- A circle whose area is 9 has a radius of $\dfrac{3}{\sqrt{\pi}}$ $(9 = \pi r^2, r = \dfrac{\sqrt{9}}{\sqrt{\pi}})$, a diameter of $\dfrac{6}{\sqrt{\pi}}$, and a circumference of $6\sqrt{\pi}$ $[C = 2\pi\left(\dfrac{3}{\sqrt{\pi}}\right)]$

The value of π is approximately 3.14 or $\dfrac{22}{7}$ (on your exam, you probably won't have to work with a value for π any more precise than "a little over 3"). In fact, in most circle problems, the solution is expressed in terms of π rather than numerically.

Circles with Triangles Inside of Them

More complex circle problems typically involve other geometric figures as well. Most common are "hybrid" problems involving circles and triangles. Two particular varieties are especially interesting (and testworthy). The first type involves a triangle with one vertex at the circle's center and the other two vertices on the circumference [both triangles in figure (A) below]. Any such triangle must be isosceles, since the sides forming the vertex at the circle's center are each equal to the circle's radius. If the angle at the circle's center is 90°, the length of the triangle's hypotenuse (chord) must be $r\sqrt{2}$, since the ratio of the triangle's sides is $1:1:\sqrt{2}$ [$\triangle ABO$ in figure (A) below]. If the angle at the circle's center happens to measure 60°, the length of the triangle's hypotenuse (chord) must be r, since the triangle is equilateral [$\triangle CDO$ in figure (A) below].

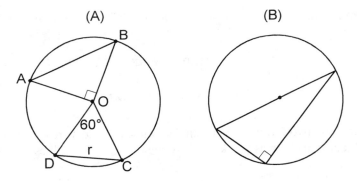

In the second variety, a triangle is inscribed inside the circle so that all three vertices lie on the circle's circumference and one side equals the circle's diameter [figure (B) above]. Any such triangle *must* be a right triangle—it must include one 90° angle. If you don't believe it, go ahead and draw some more triangles meeting these requirements (I know you will, anyway).

Square Pegs in Round Holes (and Vice Versa)

In a different type of hybrid circle problem, a square is either inscribed inside a circle [figure (A) on page 253] or circumscribed around a circle [figure (B) on page 253]. In either case, the square touches the circle at four and only four points. In figure (A), we've added two diagonals to help you see the relationship between each side of an inscribed square and the circle's radius. Each of the four triangles formed by the diagonals is a $1:1:\sqrt{2}$ triangle. In each one, the ratio of the hypotenuse (same as the side of the square) to the legs (same as circle's radius) is $\sqrt{2}:1$.

(A)   (B)

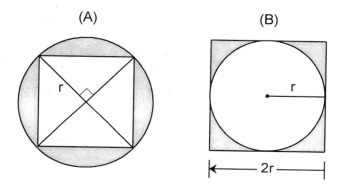

Accordingly, the area of a square inscribed in a circle is $\left(\sqrt{2r}\right)^2$ or $2r^2$. The ratio of the inscribed square's area to the circle's area is $2:\pi$. The *difference* between the two areas—i.e., the total shaded area in figure (A) above—is $\pi r^2 - 2r$. [Accordingly, the area of each crescent-shaped shaded area is $\frac{1}{4}(\pi r^2 - 2r)$.] In figure (B), however, each side of the square is $2r$ in length. Thus, the square's area is $(2r)^2$, or $4r^2$. The ratio of the square's area to that of the inscribed circle is $\frac{4}{\pi}:1$. The *difference* between the two areas—i.e., the total shaded area in the figure (B) above—is $4r^2 - \pi r^2$, or $(4 - \pi)r^2$. [Accordingly, the area of each separate (smaller) shaded area is one-fourth of that difference.]

Concentric Circles

A circle problem might involve *concentric* circles, which are two or more circles with the same center but unequal radii (creating a "bulls-eye" effect). The relationship between the areas of concentric circles depends, of course, on the relative lengths of their radii. The corresponding relationship is exponential, not linear. For example, if the larger circle's radius is *twice* that of the smaller circle's radius [as in figure (A) in the diagram below], the ratio of the circles' areas is $1:4$ $[\pi r^2 : \pi(2r)^2]$. If the larger circle's radius is *three* times the length of that of the smaller circle [as in figure (B) below], the ratio is $1:9$ $[(\pi r^2 : \pi(3r)^2]$. A $1:4$ ratio between radii results in a $1:16$ area ratio (and so forth).

(A)   (B)

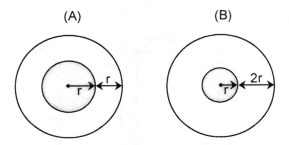

Arcs

As noted earlier, an *arc* is a segment of a circle's circumference. The size of an arc can be expressed in terms of its degree measure, its length, or both. The length of an arc (as a fraction of the circle's circumference) is directly proportionate to the arc's degree measure (as a fraction of the circle's total degree measure of 360°). This makes sense, doesn't it, since an arc and its opposite angle each establish how big a "slice" of the circle you're dealing with. Accordingly, an arc of 60° would have a length of $\frac{60}{360}$, or $\frac{1}{6}$ the circle's circumference. Given $C = 2\pi r$, that arc is $\frac{1}{6}(2\pi r)$, or $\frac{\pi r}{3}$ in length, (as illustrated by arc *AB* below):

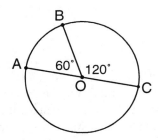

Similarly, the length of a 100° arc is $\frac{100}{360}$ $\left(\frac{5}{18}\right)$ of the circle's circumference ($2\pi r$), or $\frac{5\pi r}{9}$.

Quiz Time

Here are 10 problems to test your skill in applying the concepts in this chapter. After attempting all 10 problems, read the explanations that follow. Then go back to the chapter and review your trouble spots. If you can handle the easier problems *and* the more challenging ones, consider yourself a "smart test-taker"!

Easier

1. What is the perimeter of a square whose diagonal is 8?

2. In a parallelogram whose area is 15, the base is represented by $x + 7$ and the altitude is $x - 7$. What is the length of the parallelogram's base?

3. If the area of circle O below is 64π, what is the perimeter of the square?

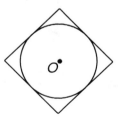

4. If a circle whose radius is x has an area of 4, what is the area of a circle whose radius is $3x$?

5. In parallelogram *ABCD* below, ∠*B* is 5 times as large as ∠*C*. What is the measure in degrees of ∠*B*?

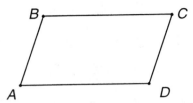

More Challenging

6. If the length and width of a rectangle are each doubled, by what percent is the area is increased?

7. Isosceles triangle *ABC* is inscribed in square *BCDE* as shown below. If the area of *BCDE* is 4, what is the perimeter of *ABC*?

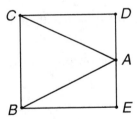

8. What is the area of trapezoid *ABCD* below?

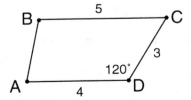

9. If the diameter of a circle is increased by 50%, by what percent is the area is increased?

10. In terms of x, what is the area of the shaded region in the figure below, which includes one circle and two squares?

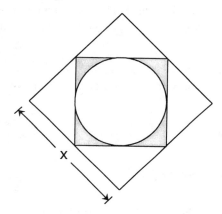

Answers and Explanations

1. The answer is $16\sqrt{2}$. The diagonal of a square is the hypotenuse of a $1:1:\sqrt{2}$ isosceles right triangle, where the two legs are sides of the square. Thus, given that the hypotenuse is 8, each side is $\dfrac{8}{\sqrt{2}}$, and the perimeter of the square is

$$4 \times \frac{8}{\sqrt{2}} = \frac{32}{\sqrt{2}} = 16\sqrt{2}.$$

2. The answer is 15. The area of a parallelogram $= (b)(h)$:

$$(x+7)(x-7) = 15$$
$$x^2 - 49 = 15$$
$$x^2 = 64$$
$$x = 8$$
$$\text{base} = x + 7 = 15$$

3. The answer is 64. The area of the circle $= 64\pi = r^2\pi$. Thus, the radius of the circle = 8. The side of the square is twice the circle's radius, or 16. Therefore, the perimeter of the square is $4 \times 16 = 64$.

4. The answer is 36. The area of a circle is πr^2. The area of a circle with a radius of x is πx^2, which is given as 4. The area of a circle with radius $3x$ is $\pi(3x)^2 = 9\pi x^2$. Therefore, the area of the larger circle is 9 times the area of the smaller circle.

5. The answer is 150. The sum of the angles in a parallelogram is 360°. Angles B and C account for half, or 180. Letting x equal the degree measure of angle C:

$$5x + x = 180$$
$$6x = 180$$
$$x = 30$$
$$\angle B = 5x = (5)(30) = 150$$

6. The answer is 300%. If the dimensions are doubled, the area is multiplied by 2^2, or 4. The new area is 4 times as great as the original area—i.e., it has been increased by 300%.

7. The answer is $2 + 2\sqrt{5}$. Each side of the square = 2. If $BE = 2$, $EA = 1$, then by the Pythagorean Theorem, BA and AC each equal $\sqrt{5}$. Thus, the perimeter of $\triangle ABC = 2 + 2\sqrt{5}$.

8. The answer is $\frac{27}{4}\sqrt{3}$. The area of a trapezoid is $\frac{1}{2}$ the product of the sum of: the two parallel sides ($BC + AD$) and the trapezoid's height. To determine its height, form a right triangle, as shown on page 259. This right triangle conforms to the 30-60-90 Pythagorean triplet. Thus, the ratio of the three sides is $1:\sqrt{3}:2$. The hypotenuse is given as 3, so the height is $\frac{3\sqrt{3}}{2}$. The trapezoid's area is as follows:

$$\frac{1}{2}(4 + 5) \times \frac{3\sqrt{3}}{2} = \frac{9}{2} \times \frac{3}{2}\sqrt{3} = \frac{27}{4}\sqrt{3}$$

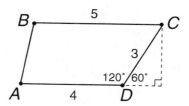

9. The answer is 125%. This question is deceptively difficult. The area of a circle is πr^2. Thus, if d (and thus r) is increased by .5, the linear ratio of the original diameter (and radius) to the larger diameter (and radius) is 1:1.5. Accordingly, the ratio of the original area to the larger area is "(1) squared : (1.5) squared," or 1:2.25. The increase is 1.25 or 125% of the original area. Confirm this by substituting a simple number such as 2 for the original radius. The area ratio is $2^2{:}3^2$, or 4:9, an increase of 125% (from 4 to 9).

10. The answer is $\dfrac{x^2(4-\pi)}{8}$. To determine the area of the shaded region, subtract the area of the circle from the area of the smaller of the two squares. First, determine the area of the smaller square. Each of the four outside triangles is a $1{:}1{:}\sqrt{2}$ right triangle, with a side of the smaller square as the hypotenuse. Each leg of these triangles is $\dfrac{x}{2}$ in length; thus, each side of the smaller square is $\dfrac{x\sqrt{2}}{2}$ in length.

Accordingly, the area of the smaller square is $\left(\dfrac{x\sqrt{2}}{2}\right)^2$, or $\dfrac{x^2}{2}$. Next, determine the area of the circle. It's diameter is $\dfrac{x\sqrt{2}}{2}$ (the length of each side of the smaller square). Thus, its radius is half that amount, or $\dfrac{x\sqrt{2}}{4}$. The circle's area is:

$$\pi\frac{x\sqrt{2}}{4}^2 = \pi\frac{2x^2}{16} = \frac{\pi x^2}{8}$$

Subtracting this area from the square's area:

$$\frac{x^2}{2} - \frac{\pi x^2}{8} = \frac{4x^2 - \pi x^2}{8} = \frac{x^2(4-\pi)}{8}.$$

15

Geometry
—Solids and the Coordinate Plane

If you understand how to determine areas of two-dimensional figures such as rectangles, triangles, and circles, you won't have any trouble handling problems involving three-dimensional objects (or "solids," as they're sometimes called in the standardized testing biz). Neither a Ph.D. in mathematics nor a pair of 3-D glasses is required! **We're going to cover** only these three basic shapes:

- rectangular solids (boxes)
- cylinders (tubes)
- pyramids

To make matters even easier, we'll only scratch the "surface" of pyramids; determining the volume of a pyramid is a bit too advanced for the basic standardized exams. Also, notice that globes (balls) and cones are missing from this short list. These shapes are a bit too tricky for basic standardized tests.

Later in the chapter we'll look at geometry problems set in the context of what's called the coordinate or *Cartesian* plane or "grid."

Rectangular Solids

A *rectangular solid* is formed by six rectangular surfaces, or *faces*, connecting at right angles at 8 corners.

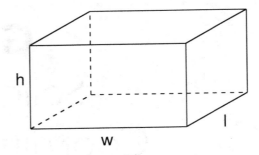

The volume (V) of any rectangular solid is the product of its three "dimensions": length, width, and height.

$$V = \text{length} \times \text{width} \times \text{height}$$

Each of three pairs of opposing faces are identical—they have the same dimensions and area. Accordingly, the surface area (SA) of any rectangular solid can be expressed as follows:

$$SA = 2(lw + wh + lh)$$

Cubes

A *cube* is a special type of rectangular solid in which all six faces, or surfaces, are square. Because all six faces of a cube are identical in dimension and area, given a length s of one of a cube's sides—or edges—its surface area (SA) is six times the square of s:

$$SA_c = 6s^2$$

Given a length s of one of a cube's sides, or edges, the *volume* of the cube is s cubed. Conversely, given the volume V of a cube, the length of one edge s is the cube root of the cube's volume V:

$$V = s^3$$

$$s = \sqrt[3]{V}$$

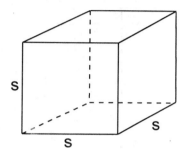

Given the area A of one face (f) of a cube, the cube's *volume* can be determined by cubing the square root of A. Conversely, given the volume V of a cube, the area can be determined by squaring the cube root of V:

$$V_c = \left(\sqrt{A_f}\right)^3$$

$$A_f = \left(\sqrt[3]{V_c}\right)^2$$

Cylinders

The figure on page 264 is a "right" circular cylinder (the tube is sliced at 90° angles). This is the only kind of cylinder you need to be concerned with. The *surface area* of a cylinder can be determined by adding together three areas: (1) the circular base, (2) the circular top, and (3) the rectangular surface around the cylinder's vertical face (visualize a rectangular label wrapped around a soup can). The area of the vertical face is the product of the circular base's circumference (i.e., the rectangle's width) and the cylinder's height. Thus, given a radius r and height h of a cylinder:

$$SA = 2\pi r^2 + (2\pi r)(h)$$

Accordingly, the surface area of the cylinder pictured below is $8\pi + 42\pi$.

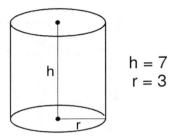

Given the radius and height of a cylinder, its *volume* can be determined by multiplying the area of its circular base by its height:

$$V_c = \pi r^2 \times h$$

Accordingly, the volume of the cylinder pictured above is $(9\pi)(7)$, or 63π .

Pyramids

Geometry problems involving pyramids are not as common as problems involving other solids; nevertheless, they are "fair game" on the exams, so you should be ready for them—just in case. Pyramids can be either 3-sided (with a triangular base) or 4-sided (with a square base). On your exam, however, expect to see only the 4-sided type. You can safely assume that all four triangular faces of a 4-sided pyramid are the same shape and size. Here's what it looks like (s is the length of one of the four sides of the square base):

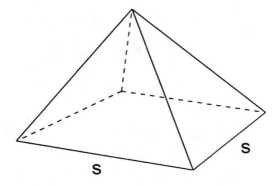

Problems involving pyramids generally involve the *altitude* and/or *surface area* of the 4-sided pyramid. The altitude is the pyramid's height—a line segment (*PQ* in the figure below) running from the pyramid's apex (highest point) down to the center of the square base.

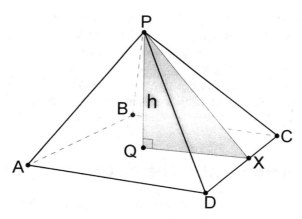

If you know the altitude and the dimensions of the square base, you can determine the area of each triangular face by applying the Pythagorean Theorem. For example, if the altitude *PQ* below is 6, and if the area of the square base is 36, then you can determine the length of *PQX* as follows:

each side of square *ABCD* is $\sqrt{36}$, or 6; thus, *QX* = 3

applying the Theorem ($c^2 = a^2 + b^2$):

$$(PX)^2 = 6^2 + 3^2$$

$$(PX)^2 = 36 + 9$$

$$PX = \sqrt{45} \text{ or } 3\sqrt{5}$$

$3\sqrt{5}$ is the "sloping" length of each of the pyramid's triangular faces. The base of each triangle is 6 (because the area of the square base is 36). Thus, the area of $\triangle PCD = \left(\frac{1}{2}\right)(6)(3\sqrt{5})$, or $9\sqrt{5}$. Accordingly, the total surface area of all four triangular sides is four times this amount, or $35\sqrt{5}$.

Coordinate Geometry

On your exam you are likely to encounter at least one or two *coordinate geometry* questions, which involve the rectangular *coordinate plane* defined by two axes—a horizontal *x-axis* and a vertical *y-axis*. Any point on the coordinate plane can be defined by two "coordinates"—an *x-coordinate* and a *y-coordinate*. A point's x-coordinate is its horizontal position on the plane, and its y-coordinate is its vertical position on the plane. The coordinates of a point are denoted by "(x,y)", where x is the point's x-coordinate and y is the point's y-coordinate.

Coordinate Signs and the Four Quadrants

The center of the coordinate plane—the intersection of the x and y axes—is called the *origin*. The coordinates of the origin are $(0,0)$. Any point along the x axis has a y-coordinate of 0 $(x,0)$, and any point along the *y*-axis has an x-coordinate of 0 $(0,y)$. The coordinate signs (positive or negative) of points lying in the four quadrants I–IV in the diagram below are as follows:

Quadrant I (+,+)

Quadrant II (–,+)

Quadrant III (–,–)

Quadrant IV (+,–)

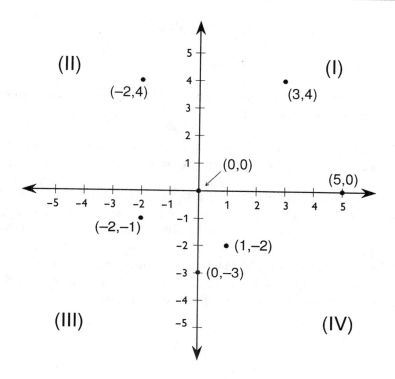

Notice in this diagram that seven different points, each with its own coordinates, have been plotted. Before you read on, make sure you understand why the coordinates of each of these points is identified as such.

Coordinate Geometry Problems

Most coordinate geometry problems involve either triangles or circles, or both. In triangle problems, your task is usually to determine the length of a sloping line segment (by forming a right triangle and applying the Pythagorean Theorem). In circle problems, your task is usually to determine the circumference or area of a circle lying on the plane. Let's look at each type.

Coordinate-Plane Triangle Problems

Consider the following two questions:

> Q1: On the coordinate plane, what is the length of a line segment with the endpoints (−2,−1) and (3,4)?

> Q2: On the coordinate plane, what is the area of a triangle whose three vertices are defined by the coordinate pairs (−2,−1) , (3,4), and (1,−2)?

The first question is easier to handle than the second one. On the coordinate plane, construct a right triangle with the line segment as the hypotenuse. The length of the horizontal leg is 5 (the horizontal distance from −2 to 3). The length of the vertical leg is also 5 (the vertical distance from −1 to 4). Thus, we are dealing with an isosceles right triangle. The ratios of the lengths of the three sides is $1:1:\sqrt{2}$. Since each leg (either of the short sides) is 5 in length, the length of the hypotenuse is $5\sqrt{2}$. The upper triangle in the following diagram illustrates the solution:

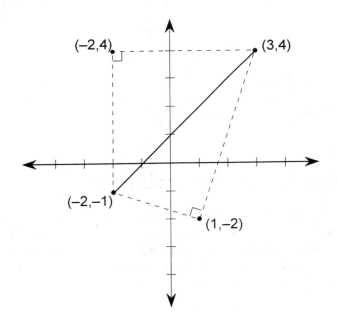

Now let's consider the second question, which establishes the lower triangle in the preceding diagram. The base and height of the triangle are represented by the dotted lines. The area of the triangle is, of course, $\frac{1}{2}bh$. So we need to determine b and h. Just as we did in the first question, think of b as the hypotenuse of a right triangle, this time with legs of 1 and 3 in length. Similarly, think of h as the hypotenuse of a right triangle whose legs are 2 and 6 in length. Do any of the convenient Pythagorean triplets allow us to shortcut applying the Pythagorean Theorem to determine each hypotenuse (the dotted line segments in the lower triangle)? No, not in either of these cases. So we need to find b and h the "long" way:

$$b^2 = 1^2 + 3^2 \qquad\qquad h^2 = 2^2 + 6^2$$
$$b^2 = 10 \qquad\qquad\qquad h^2 = 40$$
$$b = \sqrt{10} \qquad\qquad h = \sqrt{40} \text{ or } 2\sqrt{10}$$

We're not quite done. Now we need to plug these values into our formula for the area of a triangle:

$$A = \tfrac{1}{2}(\sqrt{10})(2\sqrt{10})$$
$$A = 10$$

Coordinate-Plane Circle Problems

By now you know that triangles pervade the area of geometry, and coordinate-plane circle problems are no exception. Consider, for example, the following question:

Q: On the coordinate plane, what is the area of a circle whose center is located at $(2,-1)$, if the point $(-3,3)$ lies on the circle's perimeter?

As in the previous example, construct a right triangle with the circle's radius as the hypotenuse. The length of the triangle's horizontal leg is 5 (the horizontal distance from -3 to 2), and the length of its vertical leg is 4 (the vertical distance from -1 to 3).

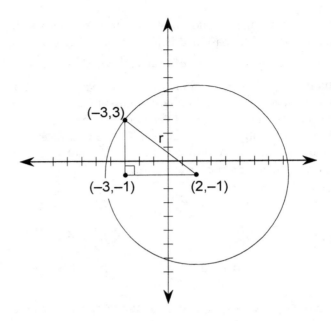

Be careful: These numbers do *not* conform to the Pythagorean triplet 3:4:5, since 4 and 5 are the lengths of the two *legs* here! Instead, you must calculate the length of the hypotenuse (radius of the circle) by applying the Pythagorean Theorem:

$$4^2 + 5^2 = r^2$$
$$16 + 25 = r^2$$
$$r^2 = 41$$
$$r = \sqrt{41}$$

Accordingly, you can now determine the area of the circle:

$$A = \pi\left(\sqrt{41}\right)^2$$
$$A = 41\pi$$

Defining a Line on the Plane Algebraically

Problems involving the algebraic equation for defining a line do *not* appear as commonly on basic math exams as the type we examined in the previous section. However, you should be ready for one—just in case; so here we go. We'll continue here to blur the "line" between geometry and algebra. Any straight line on the coordinate plane can be defined by the following algebraic equation:

$$y = mx + b$$

in which:

> m is the "slope" of the line
>
> b is the "y-intercept"
>
> x and y are the coordinates of any point on the line

Here's what all of this "slope" and "y-intercept" stuff means. Any (x,y) pair defining a point on the line can be substituted for the variables x and y in this equation. The constant b represents the line's *y-intercept* (the point on the y-axis where the line crosses that axis). The constant m represents the line's *slope*. The slope is best thought of as a fraction in which the numerator indicates the vertical change from one point to another one on the line (moving left to right) corresponding to a given horizontal change, which is indicated by the fraction's denominator. The common term used for this fraction is "rise-over-run." Keep in mind the following characteristics of certain slopes (m-values):

- a line sloping *upward* from left to right has a positive slope (m)

- a line sloping *downward* from left to right has a negative slope (m)

- a *horizontal* line has a slope of zero ($m = 0$, and $mx = 0$)

- a *vertical* line has an undefined slope (the m-term in the equation is ignored)

- a line with a slope of 1 (-1) slopes upward (downward) from left to right at a 45° angle in relation to the x-axis

- a line with a fractional slope between 0 and 1 (−1) slopes upward (downward) from left to right but at *less* than a 45° angle in relation to the *x*-axis

- a line with a slope greater than 1 (less than −1) slopes upward (downward) from left to right at *more* than a 45° angle in relation to the *x*-axis

Now consider the following question:

Q: **Which of the following points lies on a line located on the coordinate plane and having a slope of $-\frac{3}{2}$ and a y-intercept of −2?**

(A) $\left(-\frac{3}{2}, -2\right)$

(B) $(4, 6)$

(C) $\left(\frac{3}{8}, -\frac{3}{2}\right)$

(D) $\left(-\frac{8}{3}, 2\right)$

(E) $\left(-2, -\frac{3}{2}\right)$

Here's how to analyze this problem. Substitute each value pair into the equation $y = -\frac{3}{2}x - 2$. The only (x,y) pair that satisfies the equation is $\left(-\frac{8}{3}, 2\right)$. Response (D) is the correct answer. On the coordinate plane, the line appears as follows:

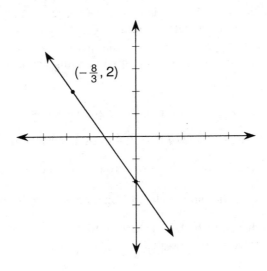

Quiz Time

Here are 10 problems to test your skill in applying the concepts in this chapter. After attempting all 10 problems, read the explanations that follow. Then go back to the chapter and review your trouble spots. If you can handle the easier problems *and* the more challenging ones, consider yourself a "smart test-taker"!

Easier

1. A rectangular box with a square base contains 24 cubic feet. If the height of the box is 18 inches, how long is each edge of the base?

2. What is the maximum number of rectangular boxes, each measuring 2" by 3" by 5", that can be packed into a rectangular packing box measuring 18" by 19" by 35", if all of the smaller boxes are aligned in the same direction?

3. A certain cylindrical pail has a diameter of 14 inches and a height of 10 inches. If there are 231 cubic inches to a gallon, which of the following is the closest approximation of the number of gallons the pail will hold?

(A) 4.8
(B) 5.1
(C) 6.7
(D) $14\frac{2}{3}$
(E) 44

4. On the coordinate plane, how far is the point (−4, −3) from the origin?

5. On the coordinate plane, what is the length of a line segment with the endpoints (−5,9) and (4,−3)?

More Challenging

6. Find the edge, in inches, of a cube whose volume is equal to the volume of a rectangular solid 2 inches by 6 inches by 18 inches.

7. If the volume of one cube is 8 times greater than that of another, what is the ratio of the area of a face of the larger cube to the area of a face of the smaller cube?

8. The volumes of two rectangular solids having the same proportions are 250 and 128. If one edge of the larger solid is 25 in length, what is the length of the corresponding edge of the smaller solid?

9. In a particular 4-sided pyramid, each side of the square base is 50 feet in length. If the apex of the pyramid is 60 feet from the ground, what is the total surface area of the pyramid, excluding the base? (Assume that all triangular faces are equal in area.)

10. On the coordinate plane, points P and Q are defined by the coordinates $(-1,0)$ and $(3,3)$, respectively, and are connected to form a chord of a circle which also lies on the plane. If the area of the circle is $\frac{25}{4}\pi$, what are the coordinates of the center of the circle?

Answers and Explanations

1. The answer is 4. The volume of a rectangular box is the product of its length, width, and height. Since the height is 18 inches, or $1\frac{1}{2}$ feet, and the length and width of the square base are the same, we can use the same variable (such as x) to represent l and w in the area formula, then solve for x:

$$x \cdot x \cdot 1\tfrac{1}{2} = 24$$
$$x^2 = 16$$
$$x = 4$$

2. The answer is 378. This question requires a bit of intuition. The objective is to minimize the unused space in the packing box by turning the smaller boxes on their appropriate sides. Align the 2″ edge of each box along the 18″ edge of the packing box (9 boxes make up a row). Align the 5″ side of each box along the 35″ edge of the packing box (7 boxes make up a row). Arranged in this manner with the 18″ by 35″ face of the packing box as the base, one "layer" of small boxes 3″ high includes 63 boxes (9×7). Given that the packing box's third dimension is 19″, 6 layers of boxes, each 3″ high, will fit into the packing box, for a total of 378 boxes. An unused layer 1″ high remains at the top of the box. (You could reverse the alignment of the 2″ and 3″ sides and arrive at the same result.)

3. The correct answer is (C). The volume of the cylindrical pail is equal to the area of its circular base multiplied by its height:

$$V = \pi r^2 h = \left(\frac{22}{7}\right)(49)(10) = 1540 \text{ cubic inches}$$

The gallon capacity of the pail $= \dfrac{1540}{231}$, or about 6.7.

4. The answer is 5. The x-coordinate is 4 units from the x-axis, while the y-coordinate is 3 units from the y-axis. These two distances are the two legs of a 3:4:5 triangle. The length of the hypotenuse, 5, is the distance of the point from the origin.

5. The answer is 15. On the coordinate plane, construct a right triangle with the line segment as the hypotenuse. The length of the horizontal leg is 9 (the horizontal distance from –5 to 4), and the length of the vertical leg is 12 (the vertical distance from –3 to 9). 9 and 12 are multiples of 3 and 4, conforming to the 3:4:5 Pythagorean triplet. Thus, without calculating the length of the length of the line segment using the Theorem, you can quickly determine the length is 15 (the Pythagorean triplet 3:4:5 is equivalent to 9:12:*15*)

6. The answer is 6. First, determine the volume of the rectangular solid:

$$V = l \cdot w \cdot h = 2 \cdot 6 \cdot 18 = 216$$

Equate this volume with the volume of the cube and solve for s (the length of any edge of the cube):

$$V = s^3$$
$$216 = s^3$$
$$6 = s$$

7. The answer is 4:1. The ratio of the two volumes is 8:1; thus, the linear ratio of the cubes' edges is the cube root of this ratio: $\sqrt[3]{8} : \sqrt{1}$, or 2:1. The area ratio is the square of the linear ratio, or 4:1.

8. The answer is 20. Since the two solids are proportionately identical, the ratio of the volumes is equal to the cube of their linear ratio. The ratio of their volumes can be expressed and simplified in this way:

$$\frac{250}{128} \text{ or } \frac{125}{64}$$

From here we can easily determine that the linear ratio of the two edges is 5 to 4:

$$\frac{\sqrt[3]{125}}{\sqrt[3]{64}} = \frac{5}{4} \text{ or } 5:4$$

Using the proportion method, set up an algebraic equation to solve for the length of the smaller edge (x):

$$\frac{5}{4} = \frac{25}{x}$$
$$5x = 100$$
$$x = 20$$

9. The answer is 6,500 square feet. The altitude of the pyramid (60) and one-half the length of a side (25) form the legs of a right triangle whose hypotenuse is the sloping (angular) height of each face. This triangle is a 5:12:13 right triangle whose sides are 25, 60, and 65. (The sloping height of each triangular face is 65.) You can now determine the area of each triangular face:

$$A = \tfrac{1}{2}(50)(65)$$

$$A = 1,625$$

Accordingly, the total surface area of the pyramid is 4 times this amount, or 6,500 square feet.

10. The answer is $(1, 1\frac{1}{2})$. Given that the area of the circle is $\frac{25\pi}{4}$, you can determine the circle's radius and diameter:

$$A = \pi r^2$$

$$\frac{25\pi}{4} = \pi r^2$$

$$\frac{25}{4} = r^2$$

$$r = \frac{5}{2}$$

$$d = 5$$

On the coordinate plane, the distance between the points whose coordinates are $(-1,0)$ and $(3,3)$ is 5 (the chord forms the hypotenuse of a 3:4:5 right triangle, as illustrated below). Because these two points are 5 units apart, chord PQ must be the circle's diameter. The circle's center lies on chord PQ midway between P and Q. The x-coordinate of the center is midway between the x-coordinates of P and Q (-1 and 3), while the y-coordinate is midway between the y-coordinates of P and Q (0 and 3). Accordingly, the center of the circle lies at the point $(1, 1\frac{1}{2})$.

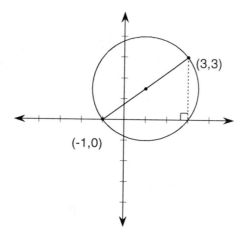

$2^2 \to 4$

$2^3 \to 8$

$24 \quad 24$

$25 \quad 96$

$\dfrac{24}{\,4\,}$
$\dfrac{}{96}$

$2^2 \to 4 \qquad 3^2 = 9$

$2^3 \to 8 \qquad 3^3 = 27$

$2^4 \to 16 \qquad 3^4 = 81$

$2^5 \to 32 \qquad 3^5 \quad 243$

$2^6 \to 64 \qquad 3^6 \quad 729$

$2^7 \to 128$

$2^8 \to 256$

$$z^2 \rightarrow 4$$
$$z^3 \rightarrow 8$$
$$24 \rightarrow 24$$